Biomarker Discovery in the Developing World: Dissecting the Pipeline for Meeting the Challenges

Sanjeeva Srivastava
Editor

Biomarker Discovery in the Developing World: Dissecting the Pipeline for Meeting the Challenges

Springer

Editor
Sanjeeva Srivastava
Department of Biosciences
and Bioengineering
IIT Bombay
Mumbai, Maharashtra, India

ISBN 978-81-322-2835-6 ISBN 978-81-322-2837-0 (eBook)
DOI 10.1007/978-81-322-2837-0

Library of Congress Control Number: 2016949744

Printed on acid-free paper

This Springer imprint is published by Springer Nature
The registered company is Springer (India) Pvt. Ltd.
The registered company address is: 7th Floor, Vijaya Building, 17 Barakhamba Road, New Delhi 110 001, India

Preface

The double burden of various communicable and non-communicable diseases in developing countries indicates the urgent need for the establishment of novel biomarkers so as to improvise the current diagnostic and therapeutic methodologies (Boutayeb and Boutayeb 2005). The advent of high-throughput technologies, which aid in the biomarker discoveries using various multi-omics platforms, has encouraged researchers globally to combat with the heterogeneity and complexities associated with life-threatening diseases to an appreciable extent. This book provides an overview of the various roadblocks that researchers face during biomarker discovery, especially in context to the unique problems posed in the developing world.

The perspective of the developing world with regard to biomarker discovery has many layers to it. At the onset, the first hurdle that is faced by researchers is the availability of well-documented and reliable biospecimens, which are representative of a disease under study. It is for this reason that practices of biobanking and establishing a network proficient in documenting, storing and sharing biospecimens must be endorsed in the developing world with due consideration of standards set worldwide. The necessity of these measures and suggestions to ease these primary hurdles is focused on in Chap. 1. The nature of the diseases studied in the developing world are often perceptible in a heterogeneous manner globally, for instance, the cancer types evident in the developed countries vary from the ones in the developing countries (Brown et al. 2006). In this light, understanding the geographic pervasiveness of diseases in the developing world context provides an opportunity for researchers, and it has been an important facet focused on in Chap. 2.

The advent of various omics approaches like genomics, transcriptomics and proteomics, which aim at identifying the arrangement, location and function of the biological entities like genes, transcripts and proteins, respectively, that constitutes a cell, has not only provided clues to understanding the geographic biasness of the diseases but also helped greatly in the development of personalised therapeutics to aid in resolving the problems associated with diseases endemic to a certain population (Cho 2010). Mass spectrometry has been central to proteomic analysis, which aids in the high-throughput analysis of diseases' pathogenesis, diagnosis and identification of therapeutic targets through discovery of potential protein and metabolite biomarkers in various diseases of the developing countries (Hawkridge and Muddiman 2009). Over the decades, the contribution of mass spectrometry in proteomics, metabolomics and tissue imaging (Aichler and Walch 2015) is tremendous and has been evolving continuously even in analysing the post-translational modifications quantitatively (Hennrich and Gavin 2015). The roles of these omics-based technological interventions, experimental structures and documentation followed by high-throughput analysis are discussed in Chap. 3. Chapter 4 focuses on mass spectrometry, which is at the forefront of proteomics-based biomarker research.

The biggest challenge in most of the researches revolving around the dreadful diseases is the paucity of adequate and sustainable financial aid for enduring research goals (Harris 2004). The current challenge cannot be tackled solely by either government or private organisations but by filling the void that exists between both. The advent of concepts like "angel philanthropy" and "crowd funding" have opened new avenues to financial abundance in the research areas, which is discussed in Chap. 5. Angel philanthropy, although initiated with higher risks for failure, aims optimistically at the great possibilities of the disruptive researches especially in cancer. These funding resources will contribute tremendously in boosting the high-risk projects in various complex diseases with less financial hindrances.

In addition to the financial encumbrance faced by the developing countries, another important aspect for consideration is the development of research and development. R&D can be established in a true sense only when there is an interactive bridge built between the scientific communities and the citizens of the country, together coequally contributing to the social innovations. Open-learning/e-learning and virtual labs are the novel initiatives which have dissolved the communication voids existing in the societies of the developing world. e- and open learning are not just encouraging distant learning in resource-limited developing countries but also have provided a platform to R&D and the social communities to help each other to strengthen the bioeconomy and social innovations, thus aiding in the development of a successful technologically sound twenty-first-century knowledge ecosystem (Ray et al. 2012). These ideas are far from the traditional ideas of propagating education and awareness but however are central in catering to the ever-increasing demands of the developing world, which is discussed in Chap. 6. An allied area growing hand in hand with biomarker discovery is the development of sustainable biosimilars. The role of proteomics in developing these biosimilars is elaborated in Chap. 7.

At the end of the biomarker discovery pipeline, an important aspect is the investigation of biomarkers by regulatory bodies to ensure translation and filing of patent. The stringent regulations implied by federal bodies often ensure only reliable biomarkers to guarantee successful translation into clinics. Therefore, along with focusing on the novel and constructive innovations, a developing country should also strive to strengthen its patent regime as it significantly contributes to innovation and technological changes in the economy (Sharma and Saxena 2012). Currently, tremendous attention is paid towards patenting novel biomarkers in both infectious and non-infectious disease, which are a major burden on the developing nations. These outputs will motivate the scientific community to carry out robust research, thus contributing to the overall growth of the country. An overview of the patent scenario in the developing world with regard to infectious and non-infectious disease is provided in Chap. 8. Considering that translation is the ultimate goal of biomarker researchers, it is of essence to be aware of the perspective and high standards set by these regulatory bodies to abide by them better, so as to positively contribute to the clinical scenario. These considerations are elaborated in Chap. 8.

Thus, this book, with the help of researchers and clinicians actively involved in translational research, holistically attempts to cover the various facets of the biomarker discovery pipeline by dissecting it into sections such as the acquisition of biospecimens and its moderation, experimental protocols and innovations in technology, funding, education and awareness, ethical and social aspects and the perspective of regulatory bodies ensuring smooth translation of the research under study, which pose as hurdles for researchers, especially in the developing world.

Mumbai, India Sanjeeva Srivastava

References

Aichler M, Walch A (2015) MALDI imaging mass spectrometry: current frontiers and perspectives in pathology research and practice. Lab Invest 95:422–431. doi:10.1038/labinvest.2014.156

Boutayeb A, Boutayeb S (2005) The burden of non communicable diseases in developing countries. Int J Equity Health 4:2. doi:10.1186/1475-9276-4-2

Brown ML, Goldie SJ, Draisma G, et al (2006) Health service interventions for cancer control in developing countries. In: Jamison DT, Breman JG, Measham AR, et al. (eds) Disease control priorities in developing countries, 2nd edn. World Bank, Washington, DC

Cho W (2010) An omics perspective on cancer research. Springer

Harris E (2004) Building scientific capacity in developing countries. EMBO Rep 5:7–11. doi:10.1038/sj.embor.7400058

Hawkridge AM, Muddiman DC (2009) Mass spectrometry-based biomarker discovery: toward a global proteome index of individuality. Annu Rev Anal Chem Palo Alto Calif 2:265–277. doi:10.1146/annurev.anchem.1.031207.112942

Hennrich ML, Gavin AC (2015) Quantitative mass spectrometry of posttranslational modifications: keys to confidence. Sci Signal 8(371):re5. doi:10.1126/scisignal.aaa6466

Ray S, Koshy NR, Reddy PJ, Srivastava S (2012) Virtual labs in proteomics: new E-learning tools. J Proteomics 75:2515–2525. doi:10.1016/j.jprot.2012.03.014

Sharma R, Saxena KK (2012) Strengthening the patent regime: benefits for developing countries-a survey. JIPR 17:122–132

Contents

About the Editor

Sanjeeva Srivastava Associate Professor, Indian Institute of Technology Bombay, India. Proteomics is truly a global science with contributors from numerous countries and continents. Dr. Srivastava is an active contributor to global proteomics science and innovation. He serves on the Executive Committee of the Proteomics Society of India, and is a member of HUPO and US-HUPO. He contributes actively to the "HUPO Website and Social Media Committee" as a member and the Human Infectious Diseases HPP initiative as a coordinator. In HUPO conferences he has participated as invited speaker, session chair, and judge. In an effort to enhance proteomics education and awareness among the community and citizens, his group has recently developed a proteomics documentary, which portrays the proteomics journey from lab to "innovation-in-society". As a Canadian trained in proteomics at Alberta and Harvard universities, he returned to India in 2009 and has established an innovative proteomics laboratory at the IITB. The research focus of Sanjeeva's team centers on using high-throughput proteomics in brain tumors and infectious diseases with over 75 publications accomplished in a span of 6 years.

He has organized two successful conferences – PSI-2014 and Targeted Proteomics International Symposium in 2015. He has published three special issues, "Proteomics in India" for Journal of Proteomics "Proteomics Research in India" for Nature India, and "Protein Arrays" for Proteomics. Having extensive teaching experience at IITB and experience of conducting proteomics courses at CSHL provided him with the background to increase proteomics education for the global community. One of his special contributions has been the development of e-learning resources (MOOC, Virtual Proteomics Laboratory). Taken together, his series of research and teaching initiatives is contributing to and fully synchronized with the mission and current actives of HUPO. He continues to develop proteomics science and innovation together with and for the next generation of keen students, researchers, and the research and education commons in Asia and global HUPO community.

Exigencies of Biomarker Research in the Developing World: A Focus on the Dearth of Biobanking Resources

1

Shabarni Gupta, Vimala Ashok Mani, Arunanshu Talukdar, Kunal Sehgal, C.S. Pramesh, Aliasgar Moiyadi, and Sanjeeva Srivastava

Abstract

The enormous burden of infectious as well as noninfectious diseases makes biomarker discovery-based research an imperative in the developing world. The extent of diversity and heterogeneity in the type of diseases that plague the low- and middle-income group nations often show a stark difference with the diseases that affect the developed countries. In order to enable global efforts to combat any given disease, it is important for researchers to have a large number of reliable biospecimens on which they could validate their findings. Inappropriate representation of samples at primary stages of research has sometimes been attributed to researchers not being able to find gold-standard biomarkers. To capture the disease heterogeneity in the subjects ranging from genetic and ethnic diversity to the underlying pathogenesis, it is important to create a resource which could facilitate the availability of biospecimens from a large cohort of subjects along with their clinical annotation, which can be studied by researchers for reliable biomarker discovery. Moreover, biorepositories are also necessary resources for facilitating validation-based studies following the discovery phase. This chapter summarizes the pressing need for

S. Gupta • V.A. Mani • S. Srivastava (✉)
Department of Biosciences and Bioengineering, Indian Institute of Technology Bombay, Powai, Mumbai 400076, Maharashtra, India
e-mail: sanjeeva@iitb.ac.in

A. Talukdar
Department of Medicine, Medical College and Hospital Kolkata, 88, College Street, Kolkata 700073, India

K. Sehgal
P. D. Hinduja Hospital and Medical Research Centre, Veer Savarkar Marg, Mahim, Mumbai 400016, Maharashtra, India

C.S. Pramesh
Department of Surgical Oncology, Tata Memorial Centre, Mumbai 400012, Maharashtra, India

A. Moiyadi
Department of Neurosurgery, Tata Memorial Centre, Mumbai 400012, Maharashtra, India

S. Srivastava (ed.), *Biomarker Discovery in the Developing World: Dissecting the Pipeline for Meeting the Challenges*, DOI 10.1007/978-81-322-2837-0_1

well-planned and managed biorepositories as one of the primary steps to facilitate reliable biomarker discovery in the developing world.

Keywords
Biobanks • Biorepositories • Developing World • Biomarker Discovery

1.1 Introduction

Interventions in healthcare research along with biopharmaceutical demand have resulted in a surge in biomarker research globally. The beginning of the new millennium also marked the beginning of the genomics era, following the completion of the Human Genome Project in 2001 (Venter et al. 2001). The post-genomics era also saw tremendous progress in other omics-related fields, where proteomics was perceived to be one of the most important tools to understand organisms, at a functional level in the coming decades. This impetus had led to the first comprehensive drafts of the Human Proteome in 2014 (Kim et al. 2014; Wilhelm et al. 2014). Thus, biosciences and healthcare research witnessed a complete face change, beginning with traditional technologies followed by the rise of genomics. This, in turn, has led to the rise of disciplines using a multi-omics approach which includes genomics, proteomics, lipidomics, metabolomics, etc. (Montague et al. 2015; Uhlén et al. 2015; Stegemann et al. 2014; Cheng et al. 2015).

Our ability to use modern technology and innovations to understand the fundamental biological basis of a disease in terms of understanding the physiology and dissecting the pathology has initiated the development of sophisticated precision medicine like never before (Collins and Varmus 2015; Roychowdhury et al. 2011). However, to cater to such wide varieties of technologies, which form the foundation of future healthcare innovations, logistic and technical support is required. Research progress in these areas relies heavily on readily available and accessible biospecimens with annotated clinical data. Unfortunately, procuring an assortment of high-quality biospecimens from large and diverse cohorts has proven to be a bottleneck in biomarker research (Hanash 2011). The allied social and ethical issues with sample procurement along with ownership are additional challenges. We discuss the need for biorepositories in biomarker research in context of the developing world.

1.2 Burden of Diseases in the Developing World

Burden of disease refers to the impact of diseases at the individual or societal levels and/or the economic costs of such diseases. An evident double burden of both communicable and noncommunicable diseases is witnessed in all the developing countries (Boutayeb and Boutayeb 2005). According to the WHO report, the mortality rates associated with the most lethal diseases has not changed significantly over the past decade. Among the non-communinicable diseases (NCD), coronary artery diseases, strokes, chronic obstructive lung disease, cancer, and diabetes stand out to be the deadliest diseases affecting developing countries with no significant decrease in the mortality and morbidity rate over the past decade (Boutayeb and Boutayeb 2005). It is estimated that in 2012, about 7.4 million deaths, accounting for 13.2 % of all the deaths, were caused by coronary artery disease. Next in the list are stroke and chronic obstructive lung diseases, responsible for 6.7 million (11.9 %) and 3.1 million (5.6 %) deaths, respectively. Approximately 1.6 million deaths are caused due to lung cancers which account for 2.9 % of deaths worldwide (WHO factsheet 2014).

Coequally, communicable diseases such as HIV AIDS, tuberculosis, acute respiratory infections, and malaria are rampant in most of the developing countries. According to the statistics reported by Foundation for AIDS Research (amfAR) in 2013, about 1.5 million deaths representing 2.7 % of the deaths globally can be attributed to AIDS (UNAIDS factsheet 2014). Another communicable disease worth mentioning is malaria, which according to the WHO contributed to 6,27,000 deaths in 2012 (WHO factsheet on the World Malaria report 2013). The next lethal infectious killer disease is tuberculosis, which contributed to 9,00,000 deaths in 2012 according to WHO reports (WHO factsheet 2014).

The heterogeneity in the spread of diseases around the various geographical locations in the globe is an issue of clinical research, according to The Global Burden of Disease Study. The likely explanation for this diversity includes ethnicity, diet, environmental risks, and infectious agents, etc., as ubiquitous risk factors. Issues more specific to the developing world include lack of access to clean water, poor sanitation, poverty and resultant lack of prioritization of healthcare, and limited access to treatment facilities (Gupta et al. 2014).

Third-world nations contribute to approximately 90 % of the global burden of diseases; however, only 10 % of global research addresses this problem (Sgaier et al. 2007). To further the damage, the rate of original research into the identification and discovery of novel biomarkers is significantly limited in developing nations. The primary reasons for this limitation are the lack of formal training in research methods, inadequate (or non-existent) protected time for research and the lack of systematic funding from governmental and non-governmental resources in such nations (Gupta et al. 2014).

1.3 What Are Biobanks

Biobanks or biorepositories are essentially a robust infrastructure for maintaining and storing biological samples with clinicopathological information for aiding scientific research (Gupta et al. 2014). The purpose of having biobanks is to essentially have a hub or a repository for an assortment of biospecimens such as serum, tissue, CSF, urine, saliva, etc., which are well annotated with clinical information including outcomes of treatment and segregated on the basis of comprehensive information of a subject's family history, health records, lifestyle, genetics, multi-omics profile, etc. (GBI Research 2012; Otlowski et al. 2010).

Biorepositories are traditionally population-based or disease-based. The advantages associated with both these types of biobanks are multifaceted. While population-based biobanks essentially serve as an excellent resource for samples in cohort studies, disease-based biobanks are a portal for biological material from diseased patients with singular focus on a particular disease. Individual population-based biobanks provide an interesting starting point for performing comparative analysis to understand the demography of various countries at a molecular, cellular, or systemic level. They could potentially help determine any underlying risk factors, prevalent single nucleotide polymorphisms (SNPs), and haplotypes to guide the emerging field of precision medicine. Apart from core biospecimens, there are examples of biobanks which store donor samples transformed into permanent cell lines (cell culture banks), DNA extracted from peripheral blood mononuclear cells (DNA-banks), pathologically altered vital tissue banks which stores frozen tissues, or cell culture samples where the disease manifests, viz., tumor sample of the tissue in which it arises. With diseases like cancer, where tumor heterogeneity and drug response seem to be an overpowering issue, biobanks storing "organoids" have emerged. Organoids are long-term cultures of primary human tissues and cancers. These overcome some of the drawbacks of tumor cell lines, i.e., they present genetic diversity and therefore serve as an excellent preclinical model to study drug response toward the mutation it presents, which is established by next-generation sequencing (NGS)-based studies on these organoids (GBI Research 2012).

1.4 Scientific Considerations

Biobanks need to operate within the scientific framework of several crucial aspects including close cooperation and inputs from a multidisciplinary team including clinicians, pathologists, biologists, researchers, technicians, data managers, etc.; standardized data collection and quality control of data; accepted technologies for preservation of biospecimens and quality control of biospecimens; updated inventories of biospecimens and clinical data; comprehensive data management; and analysis tools. Besides these important points, probably the most critical aspect is an active research community with close interaction between the biobank, clinicians, and researchers translating into some early successes even with relatively small, nonrisk-averse research projects. Developing human resource and expertise in biobanking requires two essential components: first, increased focus and investment into training, with free exchange of information between biobanks and shared best practices and innovations, and second, a clear career progression path for biobanking scientists and acceptance that biobanking is not merely an infrastructure provider. All the above mentioned points are of crucial importance if biobanking is to graduate from being just an ancillary service to a distinct scientific discipline in its own right (Harris et al. 2012).

1.5 Biobanking Facilities Available Globally

Considering that about 75 % of the burden of infectious and noninfectious diseases is localized to the developing world, it is a matter of concern that most of the world's largest biobanks and biobanking networks like the Public Population Project in Genomics (P3G), Estonian Genome Project (EGP), Singapore Tissue Network, UK Biobank, and International Genomics Consortium (IGS) come from the developed world. Moreover, biobanks which store biomaterials in collaboration with one of the aforementioned consortia are also mainly from the developed world (GBI Research 2012; Gupta et al. 2014). Thus, developing countries and the heterogeneity presented by the diseases plaguing them are not adequately represented at a global interface. As a consequence, studies on, for example, breast cancer, in the leading laboratories of the world, are limited to the samples that are made available by these biobanks which do not have any representation or collaboration with similar repositories from the developing world. Prominent examples of biobanks in the developing world like The Chinese Kadoorie Study in China, The Mexico City Prospective Study biobank, and The Indian National Biobanks cannot be scaled up to the endeavors from the West. Moreover, the nature of samples collected in these biobanks is extremely specific. For example, The Chinese Kadoorie Study and The Mexico City Prospective Study biobanks sample blood specimens of adults with noncommunicable diseases over 35 years. Thus, the need for a broader cohort of samples and a cohort of varied samples made available to researchers especially from the developing world is of prime importance in today's age (GBI Research 2012). A comprehensive list of few other prominent biobanking networks is available on various online directories and reports (Zika et al. 2010; Biobank Directory, Specimen Central [2016]).

Cancer is a disease that affects both the developed and developing world alike. The spectrum of cancer incidence is widely variant between more developed and less developed countries. For example, melanomas are far more common in the West as opposed to stomach or oral cancer typically affecting the developing world. Even within the same cancer, genetic variations are common. For example, in non-small cell lung cancer, the epidermal growth factor receptor (EGFR) mutation rates are approximately 50 % compared to about 20 % in Caucasian populations (Shi et al. 2014). In spite of the uniqueness that the developing world presents in diseases such as cancer, very few noteworthy cancer biorepositories representing such biospecimens can be named. The Cancer Center Tissue Bank; Fudan University, China; Malaysian Tissue Bank; Iran National Tumor Bank; and Bangkok Biomaterial Center, Thailand, are few among them. In India, the Tata Memorial Centre Tumor Tissue Repository (TTR) is one of the few biorepositories for sustainable cancer research to date. The Brain Biobank at National Institute of Mental Health and Neurosciences (NIMHANS),

Bangalore; Organ Retrieval Banking Organization at All India Institute of Medical Sciences (AIIMS), New Delhi; National Repository for Cell Lines at National Centre for Cell Science (NCCS), Pune; and Mycobacterial Repository at JALMA, Agra, are examples of few other biorepositories in India which cater to specific areas of research (GBI Research 2012). However, lacking is a single biorepository entity which captures samples of a larger cohort, even if it may cover a specific domain. The Tata Memorial Centre is planning to build a large biorepository to collect blood and urine samples from a large prospective cohort in the near future (personal communication, Dr. Rajesh Dikshit, Professor, Epidemiology & PI IARC Regional Hub-Mumbai, Tata Memorial Centre). With standalone biobanks, it is often observed that the samples are more often from cases which are local and do not necessarily capture the heterogeneity of an entire nation, let alone an entire continent. This is an area which has to be constructively worked on with large scale collaborations.

1.6 Hurdles in Establishing Biorepositories in the Developing World

The reasons for the lack of adequate biobanking facilities in the developing world are multifactorial. The primary reason is the lack of awareness of the importance of biobanking. In a developing world setting, where research is more often a pilot or exploratory study rather than global profiling, the sample size of clinical subjects usually ranges from about 10 to 100 on an average. Laboratories often undertake such projects in collaboration with enthusiastic clinicians who are passionate about research. However, what is lacking in such a setting is consistency in sample availability, nature of samples which may differ if it comes from two or more different clinics, similar sample processing, clinical history, and data processing, which are different in different organizations. These variations play a huge role in the outcome of OMICs-based experimental setting. On the other hand, a biorepository storing a plethora of various samples, processed in a uniform way with standard practices with patient history documented in a uniform way from the clinicians, creates a more solid foundation on which reliable experimental data can be based (Ray et al. 2013).

The second and perhaps an equally important reason is the non-availability of funding for developing such repositories. These are essentially non-profit ventures limited by the budget, as set by the government bodies to accomplish what is a massive endeavor. Thus, quality is sometimes compromised and sustenance is left largely questioned. It is therefore of importance that one plans such projects on a long-term sustainable model, so that areas of healthcare research like biomarker research is not hampered. Finally, ethical issues, governing samples, and ownership of the data taken from such repositories often clouds the development of such projects (Gupta et al. 2014).

1.7 Conclusions

Biobanks and biobanking-related activities can augment biomarker research in a major way in the developing world. However, there is a huge void that is to be filled with regard to awareness of the advantages that such platforms provide researchers and clinicians. The need does not stop at establishing new biobanks; rather, it extends to ensure its association with larger consortia, both at the national and international level. One feasible approach toward this is, to establish local biobanks which are associated with nationalized biorepositories. These national repositories should be part of a global consortium for multilateral exchange of samples and data. This will not only facilitate better data sharing among various geographic locales all over the world but will also lead to collaborative research, yielding desirable breakthroughs in life-threatening diseases (Fig. 1.1). Backing from government, philanthropists, and non-profit organizations is key to sustaining such endeavours. Fundamental to this support is the realization that biobanks provide the necessary raw material for progress in

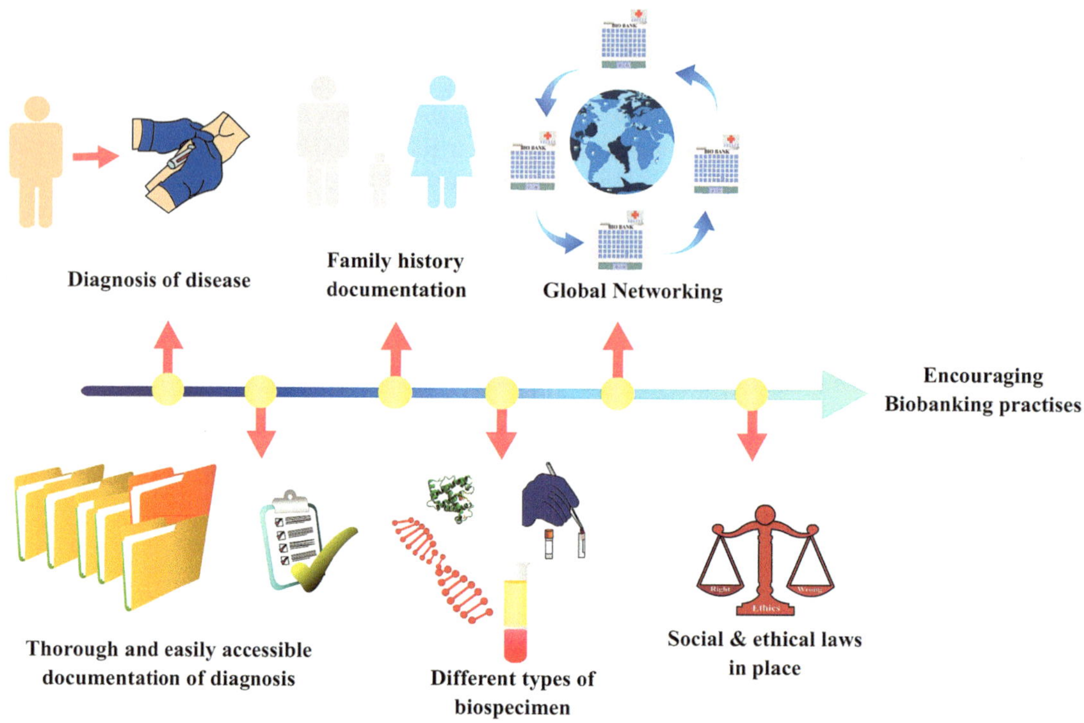

Fig. 1.1 Represents the pipeline for sustainable biobanking practices which embraces various facets involving acquisition of biospecimens, robust documentation, ethical guidelines, and global networking. This forms the foundation for biomarker discovery in the clinical scenario

scientific, medical, and biotechnology research. The path to this could be a difficult one; however, it is undeniably an essential step to sustain productive research in the field of healthcare in the future.

Acknowledgements This work has been supported by the Department of Biotechnology, Government of India, Project No. BT/PR8698/MED/29/750/2013.

Competing Interests The authors declare no competing interests.

References

Boutayeb A, Boutayeb S (2005) The burden of non communicable diseases in developing countries. Int J Equity Health 4:2. doi:10.1186/1475-9276-4-2

Cheng S, Larson MG, McCabe EL et al (2015) Distinct metabolomic signatures are associated with longevity in humans. Nat Commun. doi:10.1038/ncomms7791

Collins FS, Varmus H (2015) A new initiative on precision medicine. N Engl J Med 372:793–795. doi:10.1056/NEJMp1500523

GBI Research (2012) Biobanks—2012 year book [online], http://www.gbiresearch.com/Report.aspx?ID=Biobanks-2012-Year-Book2012

Gupta S, Venkatesh A, Ray S, Srivastava S (2014) Challenges and prospects for biomarker research: a current perspective from the developing world. Biochim Biophys Acta (5):899–908. doi:10.1016/j.bbapap.2013.12.020

Hanash SM (2011) Why have protein biomarkers not reached the clinic? Genome Med 3(10):66. doi:10.1186/gm282

Harris JR, Burton P, Knoppers BM et al (2012) Toward a roadmap in global biobanking for health. Eur J Hum Genet EJHG 20:1105–1111. doi:10.1038/ejhg.2012.96

Kim MS, Pinto SM, Getnet D et al (2014) A draft map of the human proteome. Nature 509:575–581. doi:10.1038/nature13302

Montague E, Janko E, Stanberry L et al (2015) Beyond protein expression, MOPED goes multi-omics. Nucleic Acids Res. doi:10.1093/nar/gku1175

Otlowski M, Nicol D, Stranger M (2010) Biobanks information paper. J Law Inform Sci 5:20(1)97

Ray S, Moiyadi A, Srivastava S (2013) Biorepositories for cancer research in developing countries. Nat Rev Clin Oncol 10(8):434–436. doi:10.1038/nrclinonc.2013.119

Roychowdhury S, Iyer MK, Robinson DR et al (2011) Personalized oncology through integrative high-throughput sequencing: a pilot study. Sci Transl Med 3:111ra121. doi:10.1126/scitranslmed.3003161

Sgaier SK, Jha P, Mony P, Kurpad A, Lakshmi V, Kumar R, Ganguly NK (2007) Biobanks in developing countries: needs and feasibility. Sci N Y Wash 318(5853):1074

Shi Y, Au JS-K, Thongprasert S et al (2014) A prospective, molecular epidemiology study of EGFR mutations in Asian patients with advanced non–small-cell lung cancer of adenocarcinoma histology (PIONEER). J Thorac Oncol 9:154–162. doi:10.1097/JTO.0000000000000033

Specimen Central (2016) [Online], http://specimencentral.com/biobank-directory/

Stegemann C, Pechlaner R, Willeit P et al (2014) Lipidomics profiling and risk of cardiovascular disease in the prospective population-based Bruneck study. Circulation 129:1821–1831. doi:10.1161/CIRCULATIONAHA.113.002500

Uhlén M, Fagerberg L, Hallström BM et al (2015) Tissue-based map of the human proteome. Science 347:1260419. doi:10.1126/science.1260419

Venter JC, Adams MD, Myers EW et al (2001) The sequence of the human genome. Science 291:1304–1351. doi:10.1126/science.1058040

Wilhelm M, Schlegl J, Hahne H et al (2014) Mass-spectrometry-based draft of the human proteome. Nature 509:582–587. doi:10.1038/nature13319

World Health Organization factsheet (2014) [Online], http://www.who.int/mediacentre/factsheets/fs310/en/

World Health Organization factsheet on the World Malaria Report (2013) [Online], http://www.who.int/malaria/media/world_malaria_report_2013/en/

UNAIDS factsheet (2014) [Online], http://www.unaids.org/en/resources/campaigns/2014/2014gapreport/factsheet

Zika E, Paci D, Bäumen TS, Braun A et al (2010) Biobanks in Europe: prospects for harmonisation and networking, JRC scientific and technical reports, Publications Office of the European Union. ISSN 1018-5593. doi:10.2791/41701

Geographic Pervasiveness of Cancer: Prospects of Novel Biomarker and Therapeutic Research in Developing Countries using OMICS approaches

2

Shabarni Gupta, Sandipan Ray,
Arunanshu Talukdar, Kunal Sehgal,
Aliasgar Moiyadi, and Sanjeeva Srivastava

Abstract

Lack of homogeneity in lineages of tumour cell population, in addition to constant evolution of abnormal cells, results in intratumour heterogeneity. Tumour prognosis, to a great extent, depends on the permutations and combinations in which one or more clonal lineage prevails. This directly affects therapeutic decisions as tumour sensitivity, or resistance to a particular treatment is reflected in the heterogeneity it presents. In this context, geographic predisposition presents a different dimension which may drive this heterogeneity in a particular direction and is probably the larger set of cause followed by subsets of causes like clonal evolution or cancer stem cells which result in the idiopathic nature of tumour heterogeneity. However, the geographic pervasiveness of cancers has not been studied in depth in context to tumour heterogeneity. The suffix "-omics" denotes a study in total of that particular stream of science. For example, the study of genome is referred to as genomics.

S. Gupta • S. Srivastava (✉)
Department of Biosciences and Bioengineering,
Indian Institute of Technology Bombay,
Powai, Mumbai 400076, Maharashtra, India
e-mail: sanjeeva@iitb.ac.in

S. Ray
Department of Biosciences and Bioengineering,
Indian Institute of Technology, Bombay, Powai,
Mumbai, India

Present Address: Department of Clinical Biochemistry,
Metabolic Research Laboratories, Wellcome Trust-Medical Research Council Institute of Metabolic
Science, Addenbrooke's Hospital, University of
Cambridge, Cambridge CB2 0QQ, UK
e-mail: sandipaniitbombay@gmail.com;
sr738@medschl.cam.ac.uk

A. Talukdar
Department of Medicine, Medical College and
Hospital Kolkata, 88, College Street, Kolkata
700073, India

K. Sehgal
P. D. Hinduja Hospital and Medical Research Centre,
Veer Savarkar Marg, Mahim, Mumbai
400016, Maharashtra, India

A. Moiyadi
Department of Neurosurgery, Tata Memorial Centre,
Mumbai 400012, Maharashtra, India

S. Srivastava (ed.), *Biomarker Discovery in the Developing World: Dissecting the Pipeline for Meeting the Challenges*, DOI 10.1007/978-81-322-2837-0_2

Off late, multi-omics approaches encompassing a systemic understanding of cancer tissues have shed immense light in understanding the molecular basis of tumour heterogeneity with regard to the classical models traditionally proposed to explain the heterogeneity. Genomics, proteomics, metabolomics and transcriptomics have received an enormous boost with emergence and evolution of high-throughput technology like mass spectrometry and next-generation sequencing. This chapter discusses briefly about the type of tumour heterogeneity noticed especially in context of the developing world and how OMICS-based research can bring about a revolution in understanding any geographic bias that may exist and contribute to tumour heterogeneity.

Keywords

Tumour heterogeneity • Geographic variation • Developing country • Endemic • Personalised therapeutics • Multi-OMICS

2.1 Introduction

Heterogeneity in the tumours broadly depends on the origin (tissue and cell type) of the tumour (intertumour heterogeneity) (Marusyk et al. 2012). Sometimes, despite having the same origin, tumour heterogeneity (intratumour) can be observed (Burrell et al. 2013). Intratumour heterogeneity is complex, but is a widely observed phenomenon and has been best explained using Peter Nowell's clonal evolution model and existence of cancer stem cells (Shackleton et al. 2009). Primarily, the genomic instability leading to a high genetic diversity in the tumour cell population is attributed as the primary reason or the source of heterogeneity which is driven across the clonal generations through somatic evolutions. At the same time, the non-genetic heterogeneity of deterministic and stochastic cellular phenotypes is attributed to the cancer stem cell model (Stingl and Caldas 2007; Marusyk et al. 2012). The micro environment of the tumour also has a significant influence on the epigenetic landscapes of these cells further enhancing this heterogeneity (Gupta and Massagué 2006). Furthermore, intratumour heterogeneity gets more complex in nature when a single form of clonal progeny has a pronounced dominance over the rest (Marusyk et al. 2012). There could be a situation where we have mixed dominance of two or more different clonal progenies. When such a tumour metastasises to another tissue, the heterogeneity builds up from the subtle differences depending on the differences in the tissue type between the primary tumour site and the secondary tumour site (Gupta and Massagué 2006; Bedard et al. 2013). In each of these situations, the tumour would present a different response to therapeutics and therefore poor prognosis. Heterogeneity existing in the tumour cell population exhibits some clinically important phenotypic features which renders it sensitive or resistant to certain therapies (Junttila and de Sauvage 2013), thus posing as a major hindrance in accurate diagnostic and therapeutic decision (Bedard et al. 2013). Although such phenotypic and genetic factors influencing tumour heterogeneity have been well explored, there exists a dearth of research in the field of the geographic variations which can be believed to enhance the tumour heterogeneity to another dimension.

2.2 Understanding Cancer Demographics and its Influence on Tumour Heterogeneity

A clear evidence to support the geographic pervasiveness of cancer is the fact that the cancer types that plagues the developed world is very different from the kind of cancers which burden the developing world (Rastogi et al. 2004; Gupta et al. 2014). This indicates that various pockets of populations across the globe have a predisposition towards particular cancers, and therefore conversely there are cohorts which are more "protected" from certain cancers than some other populations. If we segregate the global population into firstly two major sections—developed and developing world—there are several incidences of certain types of cancer specifically in developing nations, while there are negligible incidences for the same types of cancer in the developed world and vice versa (Gupta et al. 2014). The enormous diversities in socio-epidemiological background, ethnicity, lifestyle, diet, exposure to various environmental risk factors and infectious agents, hormonally related variables and other external epigenetic factors can be thought to contribute to cancer heterogeneity in addition to genetic factors.

Amidst this, it is also important to open up to another perspective and acknowledge the magnitude of disease burden in developing nations. While it was thought initially that low- and middle-income nations house the bulk of infectious diseases, recent observations suggest that the trends are similar even with non-communicable diseases like cardiovascular disorders, diabetes, cancer, etc. (Wagner and Brath 2012). When specifically talking about cancer, there has definitely been a rise in the cancer-incidence rate in developed world as compared to the developing world (Are et al. 2013). According to World Cancer Report 2008 (Boyle and Levin), the incidence of newly reported cancer jumped to 56 % in 2008 compared to 15 % in 1970. Out of 7.6 million cancer deaths worldwide every year, developing world accounts for two-thirds of it (Jemal et al. 2011). Hence, while genetic stability and clonal evolution-based models of studying cancer heterogeneity, post the manifestation of the disease, are the ultimate effectors of tumour heterogeneity, we propose an angle to understand cancer heterogeneity where it is more likely to first influence the kind of heterogeneity an individual is likely to present. Perhaps genetic predisposition of a population to be prone to a type of cancer is an interesting arena for -omics researchers wherein the overshadowed facet of geographic heterogeneity and its function in determining the course of detection and treatment are studied in depth.

2.3 Predisposition of a Population to a Particular Cancer

Considering that tumour heterogeneity may arise due to a large number of variables, the very root of this heterogeneity can at its onset be detected as aberrations at the genetic level. At this point, what is of interest is the likely possibility of genetic vulnerabilities lying in the population of an area endemic to a certain type of cancer and the path of heterogeneity it takes as the disease progresses as compared to any other non-endemic region. Taking India as an example of an emerging developing country, one can observe ubiquity of a particular cancer in a particular territory. It has been proven that certain gene polymorphism prevalent within a given population renders them more vulnerable to a particular cancer (Shankarkumar and Sridharan 2011). It has been observed that Indian women residing in the northern province of the country have a polymorphic copy of CYP1A1, a biomarker known to rendering them more prone to breast cancer (Singh et al. 2007), while the association of polymorphic copy of Fas-670 gene has shown to play a role in North Indian women being prone to cervical cancer (Kordi Tamandani et al. 2008). XRCC4 is known to play a key role in nonhomologous end-joining repair pathway. Thus, polymorphisms in such DNA repair genes have been shown to reduce their DNA repair capability. XRCC polymorphisms have been implicated for the prevalence of prostate cancer in North Indian Population (Mandal et al. 2011).

Several other examples within the Indian populations can be cited wherein the age-adjusted rates (AAR) of certain cancers are among the highest reported in the globe. However, the picture that has unveiled recently is that the highest AAR of a particular malignancy is focused at its epicentre and not dispersed in the entire nation uniformly. In other developing countries like China, it has been established that IL-1B2511T/T genotypes predisposes the population residing in Shanxi which has reported high prevalence of gastric cancer as compared to Guangdong which has low prevalence of gastric cancer (Zeng et al. 2003). Similarly, Korea is known to be endemic for gastric cancer (Park et al. 2000). Gall bladder carcinoma is known to be endemic in Chile (Wistuba et al. 1995), while nasopharyngeal carcinoma and oesophageal carcinoma are known to be major tumours in southern China (Seng et al. 2007). The broader question in this scenario is to find out the extent to which these cancer endemic populations are predisposed as compared to the diseased population from sporadic occurrences. The next question is to find the nature and uniqueness of these predispositions as several factors like age, sex, socioeconomic background, education, tobacco and alcohol abuse, lifestyle changes, infections (*Helicobacter pylori*, human papillomavirus, hepatitis B virus, Epstein Barr virus, etc.) and environmental factors contributing to cancer risks. However, in spite of these factors, one cannot ignore these significant demographic trends within the slow-growing pandemic of cancer which need to be studied considering these factors more holistically. Thus, these observations establish the need for specifically studying the population residing in developing nations on the basis of these trends (Nandakumar et al. 2005).

2.4 Perceiving Geographic Heterogeneity by Multi-OMICS Research

2.4.1 Genomic and Transcriptomics Studies

Due to the nature of the problem that tumour heterogeneity presents, genomics-based techniques have been traditionally used to detect signature abberations (Chung et al. 2002). RNA-/cDNA-based microarrays have been used to understand heterogeneous gene expression to understand cancer as a disease to greater depths (Alizadeh et al. 2000; Pennings et al. 2012). With the advent of next-generation sequencing, the power unleashed by genomic-based techniques has undeniably become the most important tool available to researchers to identify the players involved in tumour heterogeneity (Meyerson et al. 2010; Roychowdhury et al. 2011). In a study, a 54-year-old individual was subjected to integrated personal omics profile (iPOP). This involved a high-resolution examination of this whole genome sequence (WGS) and whole exome sequence (WES). This study provided his genomic and transcriptomic profile in addition to his proteomic, metabolomic and autoantibody profiles. These were analysed over a 14-month period through his healthy and diseased states which helped reveal certain medical risks (Chen et al. 2012). Such approaches can be extended to cancer-based cohorts as it would shed some light on tumour heterogeneity, particularly if one targets cancers in certain endemic populations.

2.4.2 Proteomic Studies

Although genomic studies are an effective method to study the heterogeneity presented in cancer, proteomic analysis of vulnerable populations can help elucidate the extent of their predisposition to a particular cancer. A comparative study of vulnerability with healthy populations can be sought out with an aim to comparatively analyse the proteome profile of relatively resistant groups to vulnerable groups (Alaiya et al. 2000). Alterations in the genome are often reflected in its proteome by several magnifications. Proteomics-based studies specially allow identification of novel biomarkers and protein expression signatures (Wulfkuhle et al. 2003). They also help in prediction of responses of a patient to a given drug, thus helping personalised therapy (Cancer Target Discovery and Development Network et al. 2010). In an attempt to reduce the invasiveness to diagnose cancer and to develop novel biomarkers, the plasma proteome has been proposed to be an attractive target to detect variation in cancer patients as compared to healthy individuals. High-throughput techniques like mass spectrometry have proved to be instrumental in understanding the differential expression of proteins in tumour tissues. Thus, such approaches provide a base to form a proteomic reference in an endemic population, and it also helps in comparing protein profiles of patients across different states within the country and in the world to explore relative risks and targeting therapy (Petricoin et al. 2002; Chumbalkar et al. 2005). The importance of such studies is to find if there are proteomic differences which may be location and race based. From the clinical perspective as well, proteomic findings can be immediately applied into translational validations, and hence the initiative to study those is in the interest of national health.

2.4.3 Metabolomic Studies

Cellular metabolism at the functional level is one of the key players, which define the physiological nature of the cell. Thereby, understanding tumour metabolism is not only crucial to understand the disease biology, but it also is important in terms of source of potential biomarkers. Metabolomics, the systemic study of metabolites, is an emerging field and is increasingly being adopted to unravel various facets of clinical oncology. Nuclear magnetic resonance (NMR), mass spectrometry (MS), Fourier-transform infrared spectroscopy (FTIR), high-performance liquid chromatography coupled with electrochemical coulometric array detection (LCECA), Raman spectroscopy and metabolite arrays are few of the technologies available to study metabolomics from various biospecimens (Kaddurah-Daouk et al. 2008; Oakman et al. 2011). Metabolomics has been used extensively and easily in biofluids where sample preparation is relatively simple; however, for tissue-based metabolomics, the sample preparation has been relatively tricky (Heinemann and Zenobi 2011). With due precaution, it is possible to study large cohorts to understand systemic alterations in biofluids like serum, urine, CSF, etc., and understand population-based metabolomics signatures. To dissect the disease completely, tissue-based metabolomics followed by single cell analysis of these tissues would be of prime importance.

2.4.4 Single Cell Analysis (SCA)

The resolution of results obtained in terms of dissecting the heterogeneity in both genomics and proteomics is a huge limiting factor. A minimum of 10^5 cells are required for extracting DNA/RNA to perform any further genomic or sequencing studies. This large number of cell input implies that they may not be single cell-derived lineages. Thus, the results from such studies may give us an averaged or holistic impression of the genomic profile (Rosenblum and Peer). Same is the case in proteomic studies wherein tissue extracts cannot be thought to be derived from cells in isolation. Single cell sequencing is an outcome of next-generation sequencing (NGS) which has helped us to now perform high-resolution screening of the single cell genome to understand the genetic, epigenetic and transcriptomic heterogeneity, especially in case of diseased samples (Macaulay and Voet 2014). Single cell proteomics in the form of flow cytometry, single cell proteomic chips, etc., have been used to decipher signalling pathways (Irish et al. 2006; Shi et al. 2012). However, there is more scope for innovation to cater to the needs for detecting proteomic aberrations at the single cell level.

Many new commercial technologies apart from the traditional high-throughput microscopy have recently emerged to take SCA to a multi-omics platform. These technologies are essentially "Lab-on-Chip" devices which exploit microfluidics, cell trapping techniques using biophysical parameters like dielectrophoretic charges, etc. (Mannello et al. 2012). Devices from ScreenCell, Fluidigm's C1™ Single-Cell AutoPrep System, Silicon Biosystems DEPArray, etc., are few such examples (Mannello et al. 2012). On culturing cell populations derived from each single cell in a given tumour tissue, it is possible to run a host of multi-omics-based experimentation (including genomics, transcriptomics, metabolomics, degradomics, etc.) on them which would truly represent the systemic alterations down to the single cell level and capture the entire biological heterogeneity. These studies can then be extrapolated at a cross-

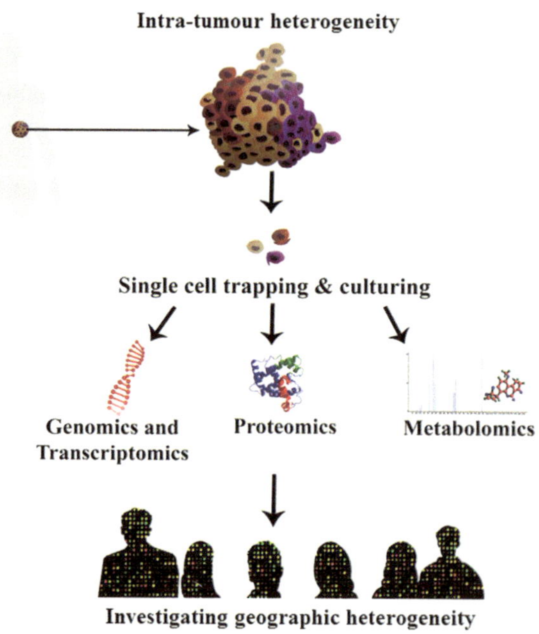

Fig. 2.1 Represents the rationale of undertaking single cell analysis to dissect intratumour heterogeneity by separating single cells in a tumour, each of which could have unique alterations of their own. These could be subjected to various multi-omics approaches like genomics, transcriptomics, proteomics, metabolomics, etc., to holistically determine the trends in a population

sectional level where cohorts from different geographical locations could be studied to find common and unique etiological alterations in cancers (Fig. 2.1).

2.5 Tumour Heterogeneity Necessitates the Personalisation of Therapeutics and Implementation of Standard Global Models

Findings of research pursued globally, primarily in the developed world, using samples from the developed countries would be futile when extrapolated to populations residing in developing nations rendering it inconclusive. Certain strategies have been explored by various research groups globally, viz., the University of Michigan in collaboration with the International Genomics Consortium and institutes contributing to The Cancer Genome Atlas (TCGA), with an aim to personalise therapy. The success of such an approach from the Broad Institute of Harvard and MIT has been recently published where pharmacological profiles of 24 anticancer drugs were studied across 479 cancer cell lines (Barretina et al. 2012). They found that the development of personalised therapeutic regimens can be enhanced by genetic predictions of drug response in the preclinical settings and their incorporation into cancer clinical trials (Barretina et al. 2012). This study is widely being used as the central rationale for most research groups towards developing personalised medication. In fact, the same institute also reported the sensitivity of Bcl-2 family antagonist, Navitoclax, resulting due to mutations in oncogene b-catenin and its candidate dependency (Basu et al. 2013). They have used this model to hypothesise novel therapeutics to accelerate the discovery of drugs to cater to patients depending on the cancer genotype and lineage. Studies affiliated to TCGA using such approaches have been responsible for major findings like classification of glioblastoma into four distinct subtypes depending on gene expression patterns and clinical characteristics, viz., proneural, neural, classical, and mesenchymal subtypes (Verhaak et al. 2010). Aberrations in expression of EGFR, NF1 and PDGFRA/IDH1 genes result in the classical, mesenchymal, and proneural subtypes, respectively (Verhaak et al. 2010). The Pan-Cancer initiative affiliated to TCGA has profiled 12 tumour types by analysing large number of human tumours to shed light on aberrations in DNA, RNA, protein or epigenetic at the molecular level (Cancer Target Discovery and Development Network et al. 2010; Kandoth et al. 2013). Their success in utilising such strategies has helped the scientific community to understand how one can extend therapies effective in one cancer type to others with a similar genomic profile (Barretina et al. 2012). On extending these strategies in collaboration with research groups in the developing world or by using samples from the developing world, the genetic basis for the disparity in global heterogeneity in cancer can be better understood.

2.6 Concluding Remarks

Tumour heterogeneity is a challenging subject to study and is essential to understand how cancer progresses and the therapeutic implications it extends. To this end, multi-omics research has provided numerous clues which establish these facts clearly. However, there is a huge scope to study geographic bias of a disease and the heterogeneity it presents. By comparing populations and the cancers endemic to these regions using multi-omics approaches, it would not only provide us biological evidences of genetic vulnerability, if it exists, but will also help extensively in certain basic generalisations of inclusions or exclusions of therapeutic regimes according to its response in a population. Further from population centric regimes, it also provides a gateway to understand the response on an individual to a particular therapy. It is probable that in the near future, technologies like NGS and Mass Spectrometry based Targeted Proteomics would be routinely used to detect residual disease after therapy during follow-ups. With the drastic evolution of expertise, personalised medicine and targeted therapeutics are very much a reality and should be pursued extensively in order to combat the challenges presented by cancer heterogeneity.

Acknowledgement This work has been supported by the Department of Biotechnology, Government of India, Project No. BT/PR4599/BRB/10/1042/2012.

Competing Interests The authors declare no competing interests.

References

Alaiya AA, Franzén B, Auer G, Linder S (2000) Cancer proteomics: from identification of novel markers to creation of artificial learning models for tumor classification. Electrophoresis 21:1210–1217. doi:10.1002/(SICI)1522-2683(20000401)21:6<1210::AID-ELPS1210>3.0.CO;2-S

Alizadeh AA, Eisen MB, Davis RE et al (2000) Distinct types of diffuse large B-cell lymphoma identified by gene expression profiling. Nature 403:503–511. doi:10.1038/35000501

Are C, Rajaram S, Are M et al (2013) A review of global cancer burden: trends, challenges, strategies, and a role for surgeons. J Surg Oncol 107:221–226. doi:10.1002/jso.23248

Barretina J, Caponigro G, Stransky N et al (2012) The cancer cell line encyclopedia enables predictive modelling of anticancer drug sensitivity. Nature 483:603–607. doi:10.1038/nature11003

Basu A, Bodycombe NE, Cheah JH et al (2013) An interactive resource to identify cancer genetic and lineage dependencies targeted by small molecules. Cell 154:1151–1161. doi:10.1016/j.cell.2013.08.003

Bedard PL, Hansen AR, Ratain MJ, Siu LL (2013) Tumour heterogeneity in the clinic. Nature 501:355–364. doi:10.1038/nature12627

Burrell RA, McGranahan N, Bartek J, Swanton C (2013) The causes and consequences of genetic heterogeneity in cancer evolution. Nature 501:338–345. doi:10.1038/nature12625

Cancer Target Discovery, Development Network, Schreiber SL, Shamji AF et al (2010) Towards patient-based cancer therapeutics. Nat Biotechnol 28:904–906. doi:10.1038/nbt0910-904

Chen R, Mias GI, Li-Pook-Than J et al (2012) Personal omics profiling reveals dynamic molecular and medical phenotypes. Cell 148:1293–1307. doi:10.1016/j.cell.2012.02.009

Chumbalkar VC, Subhashini C, Dhople VM et al (2005) Differential protein expression in human gliomas and molecular insights. Proteomics 5:1167–1177. doi:10.1002/pmic.200401202

Chung CH, Bernard PS, Perou CM (2002) Molecular portraits and the family tree of cancer. Nat Genet 32(Suppl):533–540. doi:10.1038/ng1038
Gupta GP, Massagué J (2006) Cancer metastasis: building a framework. Cell 127:679–695. doi:10.1016/j.cell.2006.11.001
Gupta S, Venkatesh A, Ray S, Srivastava S (2014) Challenges and prospects for biomarker research: a current perspective from the developing world. Biochim Biophys Acta. doi:10.1016/j.bbapap.2013.12.020
Heinemann M, Zenobi R (2011) Single cell metabolomics. Curr Opin Biotechnol 22:26–31. doi:10.1016/j.copbio.2010.09.008
Irish JM, Kotecha N, Nolan GP (2006) Mapping normal and cancer cell signalling networks: towards single-cell proteomics. Nat Rev Cancer 6:146–155. doi:10.1038/nrc1804
Jemal A, Bray F, Center MM et al (2011) Global cancer statistics. CA Cancer J Clin 61:69–90. doi:10.3322/caac.20107
Junttila MR, de Sauvage FJ (2013) Influence of tumour micro-environment heterogeneity on therapeutic response. Nature 501:346–354. doi:10.1038/nature12626
Kaddurah-Daouk R, Kristal BS, Weinshilboum RM (2008) Metabolomics: a global biochemical approach to drug response and disease. Annu Rev Pharmacol Toxicol 48:653–683. doi:10.1146/annurev.pharmtox.48.113006.094715
Kandoth C, McLellan MD, Vandin F et al (2013) Mutational landscape and significance across 12 major cancer types. Nature 502:333–339. doi:10.1038/nature12634
Kordi Tamandani DM, Sobti RC, Shekari M (2008) Association of Fas-670 gene polymorphism with risk of cervical cancer in north Indian population. Clin Exp Obstet Gynecol 35:183–186
Macaulay IC, Voet T (2014) Single cell genomics: advances and future perspectives. PLoS Genet 10:e1004126. doi:10.1371/journal.pgen.1004126
Mandal RK, Singh V, Kapoor R, Mittal RD (2011) Do polymorphisms in XRCC4 influence prostate cancer susceptibility in north Indian population? Biomark Biochem Indic Expo Response Susceptibility Chem 16:236–242. doi:10.3109/1354750X.2010.547599
Mannello F, Ligi D, Magnani M (2012) Deciphering the single-cell omic: innovative application for translational medicine. Expert Rev Proteomics 9:635–648. doi:10.1586/epr.12.61
Marusyk A, Almendro V, Polyak K (2012) Intra-tumour heterogeneity: a looking glass for cancer? Nat Rev Cancer 12:323–334. doi:10.1038/nrc3261
Meyerson M, Gabriel S, Getz G (2010) Advances in understanding cancer genomes through second-generation sequencing. Nat Rev Genet 11:685–696. doi:10.1038/nrg2841
Nandakumar A, Gupta PC, Gangadharan P et al (2005) Geographic pathology revisited: development of an atlas of cancer in India. Int J Cancer J Int Cancer 116:740–754. doi:10.1002/ijc.21109
Oakman C, Tenori L, Biganzoli L et al (2011) Uncovering the metabolomic fingerprint of breast cancer. Int J Biochem Cell Biol 43:1010–1020. doi:10.1016/j.biocel.2010.05.001
Park YJ, Shin KH, Park JG (2000) Risk of gastric cancer in hereditary nonpolyposis colorectal cancer in Korea. Clin Cancer Res Off J Am Assoc Cancer Res 6:2994–2998
Pennings JLA, Van Dycke KCG, van Oostrom CTM et al (2012) Biomarker discovery using a comparative omics approach in a mouse model developing heterogeneous mammary cancer subtypes. Proteomics 12:2149–2157. doi:10.1002/pmic.201100497
Petricoin EF, Zoon KC, Kohn EC et al (2002) Clinical proteomics: translating benchside promise into bedside reality. Nat Rev Drug Discov 1:683–695. doi:10.1038/nrd891
Rastogi T, Hildesheim A, Sinha R (2004) Opportunities for cancer epidemiology in developing countries. Nat Rev Cancer 4:909–917. doi:10.1038/nrc1475
Rosenblum D, Peer D (n.d.) Omics-based nanomedicine: the future of personalized oncology. Cancer Lett. doi:10.1016/j.canlet.2013.07.029
Roychowdhury S, Iyer MK, Robinson DR et al (2011) Personalized oncology through integrative high-throughput sequencing: a pilot study. Sci Transl Med 3:111ra121. doi:10.1126/scitranslmed.3003161
Seng TJ, Low JSW, Li H et al (2007) The major 8p22 tumor suppressor DLC1 is frequently silenced by methylation in both endemic and sporadic nasopharyngeal, esophageal, and cervical carcinomas, and inhibits tumor cell colony formation. Oncogene 26:934–944. doi:10.1038/sj.onc.1209839
Shackleton M, Quintana E, Fearon ER, Morrison SJ (2009) Heterogeneity in cancer: cancer stem cells versus clonal evolution. Cell 138:822–829. doi:10.1016/j.cell.2009.08.017
Shankarkumar U, Sridharan B (2011) Glioma Indian scenario: is there a human leucocyte antigen association? J Nat Sci Biol Med 2:205–208. doi:10.4103/0976-9668.92323
Shi Q, Qin L, Wei W et al (2012) Single-cell proteomic chip for profiling intracellular signaling pathways in single tumor cells. Proc Natl Acad Sci U S A 109:419–424. doi:10.1073/pnas.1110865109
Singh N, Mitra AK, Garg VK et al (2007) Association of CYP1A1 polymorphisms with breast cancer in North Indian women. Oncol Res 16:587–597
Stingl J, Caldas C (2007) Molecular heterogeneity of breast carcinomas and the cancer stem cell hypothesis. Nat Rev Cancer 7:791–799. doi:10.1038/nrc2212
Verhaak RGW, Hoadley KA, Purdom E et al (2010) Integrated genomic analysis identifies clinically relevant subtypes of glioblastoma characterized by

abnormalities in PDGFRA, IDH1, EGFR, and NF1. Cancer Cell 17:98–110. doi:10.1016/j.ccr.2009.12.020

Wagner K-H, Brath H (2012) A global view on the development of non communicable diseases. Prev Med 54(Suppl):S38–S41. doi:10.1016/j.ypmed.2011.11.012

Wistuba II, Sugio K, Hung J et al (1995) Allele-specific mutations involved in the pathogenesis of endemic gallbladder carcinoma in Chile. Cancer Res 55:2511–2515

Wulfkuhle JD, Liotta LA, Petricoin EF (2003) Proteomic applications for the early detection of cancer. Nat Rev Cancer 3:267–275. doi:10.1038/nrc1043

Zeng Z-R, Hu P-J, Hu S et al (2003) Association of interleukin 1B gene polymorphism and gastric cancers in high and low prevalence regions in China. Gut 52:1684–1689

Omics: Data Processing and Analysis

3

Saicharan Ghantasala, Shabarni Gupta, Vimala Ashok Mani, Vineeta Rai, Tumpa Raj Das, Panga Jaipal Reddy, and Veenita Grover Shah

Abstract

The innovations in genome sequencing technologies have emanated in better understanding of biosystems leading to the dawn of the "omics" era. Proteomics has been an integral interface in the post-genomic era, and has allowed researchers to explore other omics-based platforms like metabolomics, transcriptomics, phenomics, etc. In pursuit of obtaining a systemic understanding of biosystems, the scientific community is now largely incorporating a multi-omics-based workflow, with genomics and proteomics at the centre of this integrated approach. Techniques such as gel-based proteomics, mass spectrometry, protein microarrays and label-free platforms have emerged as powerful tools for high-throughput screening and discovery-based studies in many of these multi-omics disciplines. However, with increased throughput, large amount of data is generated, and analysis of huge data often poses a challenge to researchers. The automation in specialized software has been immensely helpful to researchers in data acquisition; however, the downstream workflow of these sophisticated technologies continues to disconcert scientists, embracing an integrated multi-omics approach. This chapter aims at providing an overview of various proteomics-based technologies and their data evaluation strategies in context to biological studies. Data storage in specialized databases also requires attention, but is beyond the scope of this chapter. Gel-based

S. Ghantasala • S. Gupta • V.A. Mani • V. Rai
T.R. Das • P.J. Reddy • V.G. Shah (✉)
Department of Biosciences and Bioengineering,
Indian Institute of Technology Bombay,
Powai, Mumbai 400076, Maharashtra, India
e-mail: veenita7shah@gmail.com

S. Srivastava (ed.), *Biomarker Discovery in the Developing World: Dissecting the Pipeline for Meeting the Challenges*, DOI 10.1007/978-81-322-2837-0_3

proteomics, mass spectrometry, protein microarrays and label-free technologies are some of the commonly employed techniques in metabolomics, interactomics, genomics and transcriptomics, thus encompassing a multi-omics perspective on data analysis.

Keywords
Omics • Data analysis • 2D-DIGE • MALDI-TOF • LC-MS-MS • Protein microarray • Surface Plasmon Resonance

3.1 Introduction

Omics refers to a field of biology aimed at elucidating the structure, function and dynamics of various biological entities that constitute a cell. Various omics technologies are used in different disciplines such as genomics, "the study of genes"; transcriptomics, "the study of expression and spatial distribution of gene at the mRNA level"; proteomics, "the study of gene function at the protein level"; and metabolomics, "the study of metabolites". The common objective of these omics approaches is the creation of large comprehensive data sets, which will help in a better understanding of the biology of organisms. Until a few decades ago, Sanger's sequencing was the sole method to gain an insight on primary structure of proteins, but recent advances in electrophoresis and chromatography, coupled with improvements in mass spectrometry, allow a better understanding of proteomes. Proteomics is the comprehensive analysis of protein structure, function and dynamics studied by investigating the protein abundance, modifications, interacting partners and pathways, in order to understand the cellular processes (Chandramouli and Qian 2009). Some of the sophisticated and high-throughput omics techniques, described in this chapter, such as 2-DE, DIGE, mass spectrometry, protein microarrays and surface plasmon resonance, make the quest of knowledge seem never ending. An overall schematic representing a typical proteomic analysis workflow using various omics tools is shown in Fig. 3.1.

With the advent of technology in proteomics, the data generation is fast, enormous and exploding in terms of size and complexity. The prime challenge today is to handle large data sets generated across the world and to analyse the same for relevant biological interpretations (Gomez-Cabrero et al. 2014). The accessibility of omics data to researchers is another matter of concern. Many proteomics data repositories have been established to safely store the vast data generated by researchers across the globe. Some of the known 2-DE data repositories include SWISS-2DPAGE, WORLD-2DPAGE Portal, EcoproDB, 2Dbase, GELBANK and proteome 2D-PAGE database. Recently, some mass spectrometry (MS)-based data repositories (PRIDE, GPMDB, PeptideAtlas, Tranche and NCBI peptidome) have also been established with generation of high-confidence data and coverage.

Besides proteomics, many other omics approaches have attained greater interest in research to offer a better understanding of biological systems. The emerging omics technologies such as transcriptomics, metabolomics and phenomics provide a critical insight into the origin, function and regulation of molecules and their pathways. With the influx of omics data, there is a growing need to know the subject better in order to combat the challenges in data management and analysis. Systems biology has undoubtedly evolved to manage, organize and process multi-omics data, in a methodical manner, by generating tools capable of analysing huge data sets (Gehlenborg et al. 1998). The current chapter

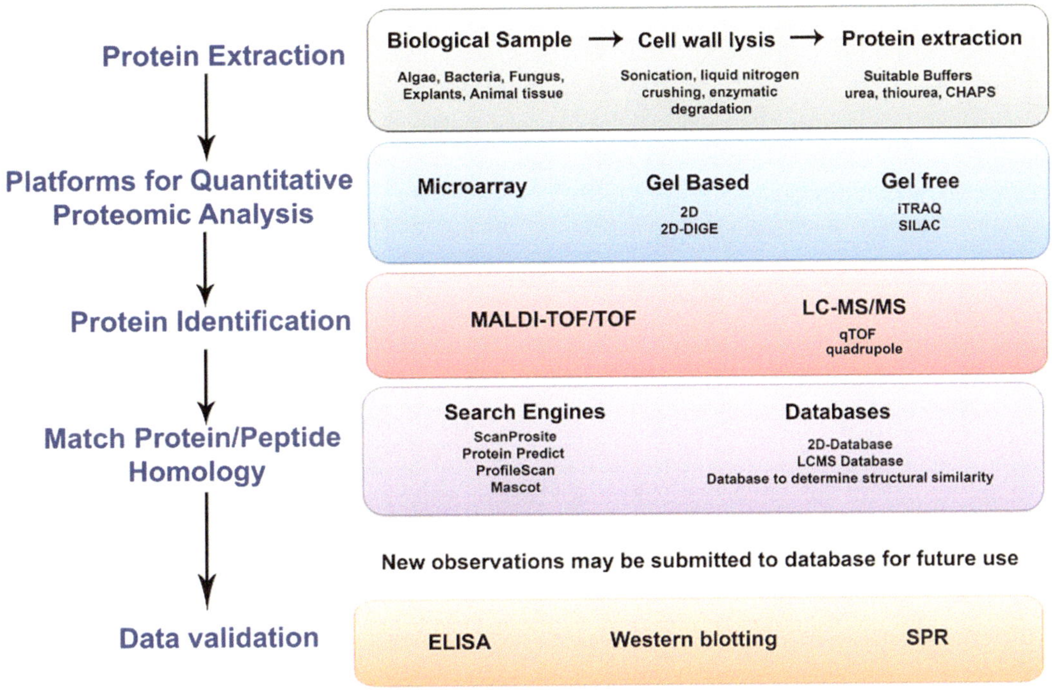

Fig. 3.1 Overall schematic representing a typical proteomic analysis workflow using various omics tools

focuses on providing an overview on some of the omics approaches, mentioned above, and their data evaluation strategies.

3.2 Two-Dimensional Gel Electrophoresis

Two-dimensional gel electrophoresis (2-DE) has been an effective gel-based method for global and quantitative proteomic analysis of various biospecimens. With its advent in the 1970s, this technique has paved the way for easy separation of complex protein mixtures, which was not possible through one-dimensional electrophoresis. The technique gained popularity in the 1990s with the introduction of immobilized pH gradient (IPG) strips which eased the protocols and increased its reproducibility and efficiency (Palzkill 2002; Beckett 2012). The basic principle involves the separation of a complex mixture of proteins based on two independent parameters: isoelectric point (pI) and molecular weight (Fig. 3.2a). The first-dimensional separation occurs through isoelectric focusing (IEF), a technique in which separation of proteins occurs based on their respective pI (the pH at which proteins form zwitterions), followed by a second-dimensional separation which is based on their molecular weight, similar to one-dimensional gel electrophoresis (1-DE). 2-DE approach can be used for global as well as differential protein profiling from biological samples (Fig. 3.2b). In the past, protein spot detection involved visual analysis of coomassie or silver-stained spots. Over the years, several software have been developed to enhance the sensitivity of protein spot detection (Table 3.1). The protein spots of interest (all spots in case of global proteomics and statistically significant spots in case of quantitative proteomics) can be excised, subjected to in-gel digestion followed by identification using mass spectrometry. The major advantage of 2-DE over 1-DE is the fact that it enables high-resolution separation of proteins with similar molecular weights but different pIs. Thus, 2-DE can be regarded as a milestone in

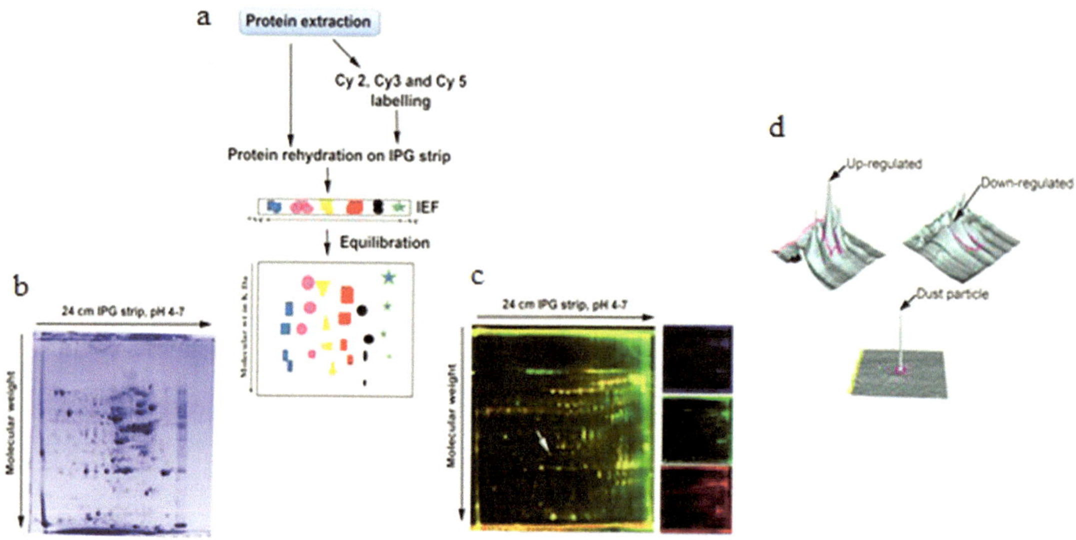

Fig. 3.2 Proteomic analysis using classical 2-DE and 2D-DIGE. A classical workflow demonstrating (**a**) 2-DE and 2D-DIGE experiment generating (**b**) coomassie-stained gel image of a complex protein mixture after classical 2-DE and (**c**) scanned image of a gel following 2D-DIGE with fluorescently labelled proteins as visible spots. 3D views for different (**d**) protein spots and dust particle as detected using DeCyder software

Table 3.1 List of commonly available software for 2-DE and 2D-DIGE analysis

S. No.	Software	Source	Applications
1	IMP7	www.gehealthcare.com	2-DE
2	DeCyder	www.gehealthcare.com	2D-DIGE
3	PDQuest	www.bio-rad.com	2-DE
4	Delta2D	www.decodon.com	2-DE and 2D-DIGE
5	GelScape	www.gelscape.ualberta.ca	2-DE
6	Flicker	http://open2dprot.sourceforge.net/Flicker	1-DE and 2-DE
7	REDFIN	www.ludesi.com/redfin	2-DE
8	Melanie	http://world-2dpage.expasy.org/melanie/	2-DE
9	ImageMaster	www.apbiotech.com	2-DE
10	ProteomeWeaver	www.defeniens-imaging.com	2-DE
11	Z3 2D-gel analysis system	www.compugen.co.il	2-DE
12	Progenesis and Phoretix 2D	www.nonlinear.com	1-DE and 2-DE

the proteomics era due to its ability to provide relatively higher protein coverage.

The limitations associated with 2-DE such as gel-to-gel variability and poor sensitivity led to the development of a relatively new technique with the ability to label proteins, which are then subjected to co-electrophoresis on a single gel. This technique, designed by Jon Menden's group, is referred to as two-dimensional difference gel electrophoresis (2D-DIGE) (Unlü et al. 1997). The methodology is similar to the conventional 2-DE except for the pre-electrophoretic labelling of samples with cyanine dyes (Fig. 3.2c). The cyanine dyes (Cy2, Cy3 and Cy5) bearing N-hydroxysuccinimidyl ester groups bind covalently with ε-amino groups of lysine residues in the protein of interest (Wilkins 2008). Each of these Cy dyes is matched for mass and charge, and is spectrally resolvable due to different excitation/emission wavelengths. Cy2 (excitation, 492 nm; emission, 510 nm) is preferably used to label the internal control, which is a pool of equal amounts of all samples loaded on the gel, whereas the samples under study (test and control) are labelled with Cy3 (excitation, 550 nm; emission, 570 nm) and Cy5 (excitation, 650 nm; emission, 670 nm) dyes. The internal control helps to eliminate the gel-to-gel variations. DIGE gels are usually run in biological replicates for gel consistency and better statistical analysis of protein spots.

3.2.1 Gel Analysis and Data Assimilation

The 2D-DIGE gels are scanned in a specialized variable mode laser scanner that enables visualization of a wide range of fluorescent wavelengths. The scanned images are further analysed using various software, some of which are enlisted in Table 3.1. These software align two or more gels in the same orientation, assign spot numbers to all protein spots in a gel and then overlay the gels. The differential expression (up- or down-regulation) in protein spots is enlisted based on their spot intensities. For instance, the DeCyder software provides options like differential in-gel analysis (DIA) for the analysis of an individual gel image and biological variation analysis (BVA) for the analysis of multiple gels (biological replicates) (GE Healthcare DeCyder 2D Software GE Healthcare 2-DE Principles and Methods 2004). DIA provides three-dimensional view of each protein along with its maximum slope and volume which aids in better understanding of the protein expression level in a given sample (Fig. 3.2d). Care must be taken to ensure meticulous gel cropping and overlaying to avoid elimination or mismatch of any protein spot. It is necessary to include certain filters during image analysis which aid in excluding some artifacts as the software cannot distinguish between protein spots and dust particles resulting in false 3D images (Fig. 3.2d). Additionally, manual curation is preferred to avoid false results and mismatching of protein spots. For the protein identification of 2D-DIGE analysis, a preparatory 2-DE gel run is preferred. However, a 2-DE gel may not be an exact replica of the corresponding DIGE gel making spot cutting a difficult task (Baggerman et al. 2005).

Owing to limitations such as limited protein coverage, higher sample requirement, low sensitivity in protein identification, low solubility of membrane-associated proteins, limited sample loading capacity of IPG strips and gel-to-gel variability, gel-based approaches (2-DE and 2D-DIGE) are increasingly being replaced by gel-free methods like iTRAQ (isobaric tag for relative and absolute quantitation) and SILAC (stable isotope labelling by amino acids in cell culture). These approaches directly label the peptides, which can be detected using mass spectrometry.

For many years, protein identification relied on a laborious technique called Edman degradation. However, there were several disadvantages linked to this method: (a) very slow and exhaustive process (only ten residues identified in 24 h), (b) required large amount of protein samples, (c) could not be performed if N-terminus of the protein was inaccessible (folded or modified) and (d) the reduction in efficiency after 50–60 residues.

The advent of high-throughput mass spectrometric technologies eased the task of protein identification and quantification with greater efficiency and accuracy. An overview of mass spectrometric analysis of proteins following the 2-DE procedure is described in the following sections.

3.3 Matrix-Assisted Laser Desorption/Ionization Time of Flight

Mass spectrometry is one of the key platforms in the field of proteomics. Out of the various mass spectrometric techniques available, matrix-assisted laser desorption/ionization time of flight (MALDI-TOF) gained popularity due to its ease of application for mass determination and protein identification. MALDI-TOF is a versatile approach for analysing proteins, peptides, oligonucleotides, glycans, polymers and organometallics, utilizing minimal reagents and easily accessible protocols.

MALDI-TOF like other mass spectrometric instruments consists of three functional components: an ion source to ionize and transfer analyte ions to gaseous phase, a mass analyser to separate ions based on their mass-to-charge ratio (m/z) and a detector to detect the separated analyte ions. Ionization techniques like MALDI and electrospray are soft-ionization approaches in mass spectrometry that allow large non-volatile biomolecules like proteins to ionize and vaporize readily (Croxatto et al. 2012).

MALDI is based on rapid photo-volatilization of analytes embedded in the matrix. A matrix is a compound capable of acquiring energy generated by laser and passing this energy to analyte molecules, thereby facilitating their desorption and ionization (Marvin et al. 2003). The excited matrix causes protonation of analytes resulting in formation of singly or doubly charged (sometimes multiply charged) ions in the gaseous phase. An electrostatic field further accelerates all analyte ions to attain equal kinetic energy. These ions further travel ahead in the flight tube to get separated by the analyser based only on their m/z ratio such that the ions with low m/z ratios travel faster compared to the ones with higher m/z. The separated ions are detected by the detector, which records the intensity of ions and generates a plot of m/z to relative ion intensity/abundance referred to as the mass spectrum. MALDI offers both positive and negative modes that can be selected during analysis depending on the charge attained by the analyte during ionization. Proteins and peptides are generally analysed using positive mode, whereas nucleic acids are analysed using negative mode.

The selection of matrix is a crucial step in MALDI, and depends on the type of analyte and objective of analysis. Preparation of matrix solution involves dissolving the solid matrix in suitable organic solvents. Pure proteins and in-gel (trypsin or other suitable enzymes) digested proteins are analysed for mass determination and protein identification, respectively. Desalting is an important step performed to remove salts and other contaminants which may result in unwanted peaks and noise in the mass spectrum. Semi-purified proteins may generate large number of peaks causing uncertainty in the data obtained.

3.3.1 Data Acquisition and Analysis

Data acquisition refers to storing of electrical signals as mass spectrum after the ions are detected by the detector. The matrix and samples are loaded on the MALDI target resulting in sample crystallization followed by data acquisition by laser bombardment (Fig. 3.3a). In the generated mass spectrum, we may often observe noisy peaks having poor signal-to-noise ratio (S/N) and higher baseline for low m/z values. Data processing involves improving the S/N ratio of peaks and correcting the baseline by base smoothening and baseline subtraction, respectively (Fig. 3.3b). This is followed by peak picking (selection of suitable peaks), which is generally performed using algorithms such as SNAP (sophisticated numerical annotation procedure) and centroid. A modification of peak picking method called two-Gaussian algorithm is utilized for proteins showing low intensity in MS spectrum due to their poor expression (Kempka et al. 2004). Kernel matching pursuit (KMP) classifier is another novel algorithm that has been used to identify differentially expressed proteins in tissues of healthy controls and patients with lung cancer (Liu et al. 2003).

Bovine serum albumin (BSA) is extensively used as a control for MALDI studies. The molecular weight of BSA is determined in linear positive mode using sinapinic acid matrix. In linear mode, generally used for labile or high molecular weight molecules (proteins), the flight tube is straight with a detector at its end. In case of BSA, two major peaks are observed in the mass spectrum after initial processing of the raw data (smoothening and baseline correction). The peak with maximum intensity corresponds to the mass of intact singly charged BSA ion ~66.5 KDa (m/z value), whereas the smaller peak ~33 KDa denotes the doubly charged BSA ion (m/2Z) (Fig. 3.4a).

Mass determination for complex samples such as polysaccharides, polymers, glycans and

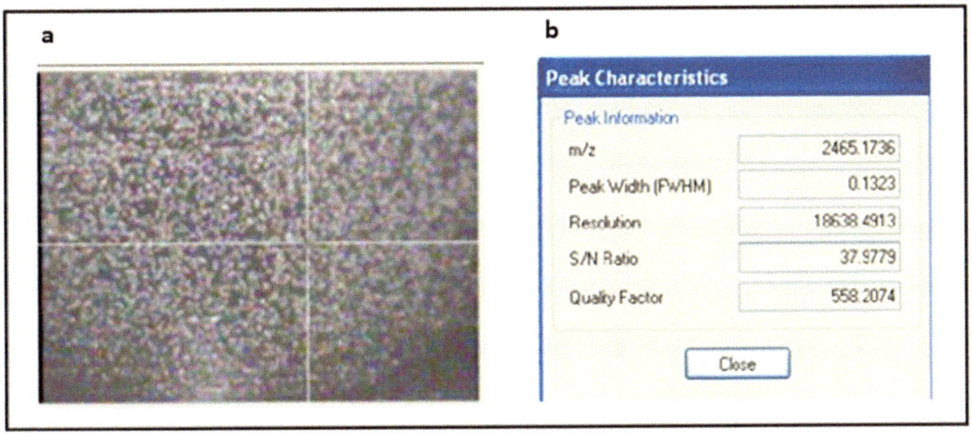

Fig. 3.3 Spotting and peak generation using MALDI-TOF. The representative figure illustrates (**a**) peptide-matrix crystals after spotting and (**b**) common peak characteristics obtained after peak generation

nucleic acids is more challenging compared to proteins. Polysaccharides (due to their high molecular weight) require additional salts and high concentration of matrix for better ionization and desorption. The MALDI analysis for oligosaccharides and polysaccharides may differ from each other due to their varying molecular weights and non-uniformity in their structure (Hao et al. 1998). For synthetic polymers, mass determination has always been questionable due to their molecular weight, polydisperse nature and the presence of additional end groups in their structure (Christian and Jackson 1997). However, MALDI-TOF plays a significant role in combating the challenges associated with mass determination of such complex samples.

For protein identification, the digested protein yields a mixture of peptides unique to the particular protein, which results in a peptide mass fingerprint (PMF) or peptide mass map. This PMF is further searched against different protein sequence databases identifying the correct match with the highest score (Webster and Oxley 2012). For BSA identification, after trypsin digestion, α-cyano-4-hydroxycinnamic acid (CHCA) is used as matrix to spot the BSA tryptic digest. The reflectron mode, designed for peptides and other small molecules, is used in such situations. In this mode, a reflectron (series of evenly spaced electrodes) is introduced in the flight tube at the end of the analyser to increase the flight length of the ions. Electrical field is applied over the reflectron so that the ions entering the reflectron undergo a continuous deceleration till they stop and leave the reflectron, where they get accelerated again to reach the detector. The reflectron thus facilitates in improving the resolution of small molecules by increasing their flight length.

The BSA peptide mass fingerprint shows many signature peptide peaks of different molecular masses: 927.4, 1,439.7, 1,479, 1,567.7, 1,724.7 and 2,044.9 (Fig. 3.4b). After processing the raw PMF, few peptide peaks with higher signal-to-noise ratios are selected for an extended step, the tandem mass spectrometry (MS-MS) (Fig. 3.4c). In tandem mass spectrometry, an additional analyser is incorporated, as observed in TOF/TOF with two TOF analysers or Q-TOF with one quadrupole and one TOF analyser. The two mass analysers in these instruments are separated by a collision cell and an ion deflection gate. The precursor ions, after being separated by the first analyser, get further fragmented in the collision cell, and the resultant daughter ions are analysed by the second analyser (Hoffman and Stroobant 2007). Tandem mass spectrometry thereby significantly increases the sensitivity of protein identification.

The MS-MS spectra for BSA with collective parent and daughter ions can then be searched

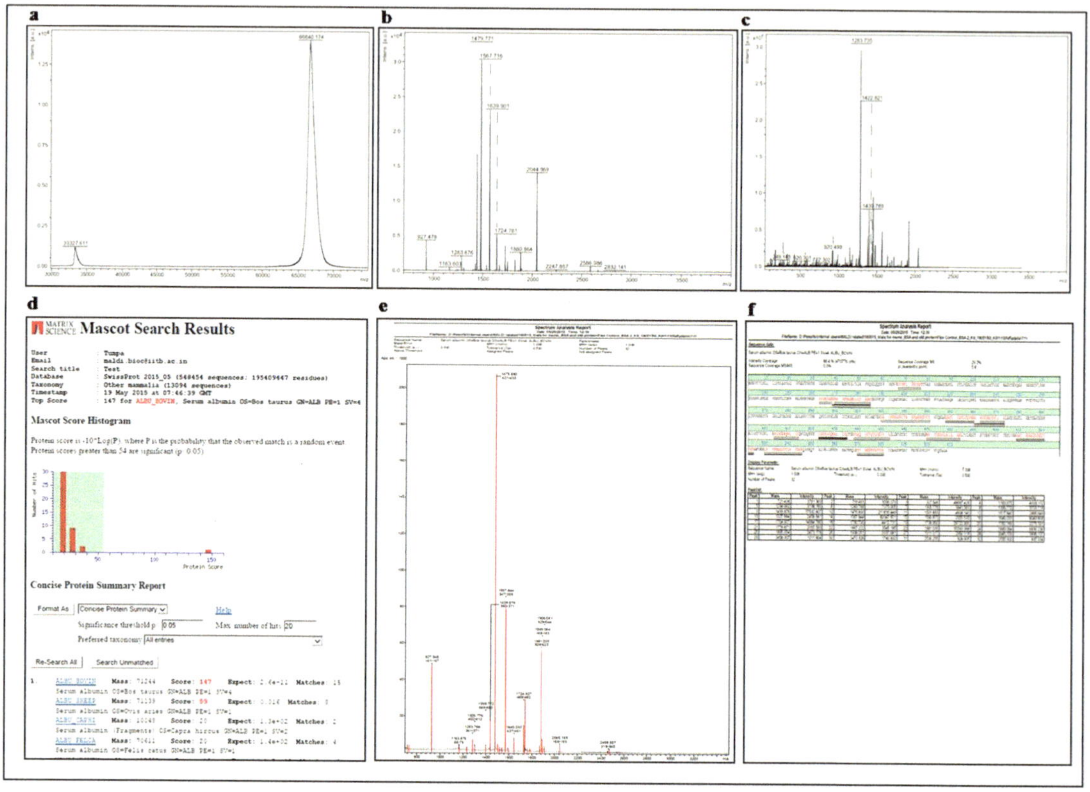

Fig. 3.4 MALDI-TOF data analysis for mass determination and identification of bovine serum albumin. The panels demonstrate (**a**) mass spectrum of undigested BSA, (**b**) peptide mass fingerprint of trypsin-digested BSA and (**c**) MS/MS spectrum of BSA. Protein identification is performed using Swiss-Prot database to derive (**d**) Mascot search results for generating (**e**) Spectrum analysis report showing (**f**) BSA sequence data

against one of the various protein databases using mascot search engine (Fig. 3.4d). The search result also displays other serum albumins from closer taxonomical groups, due to protein sequence similarity in related species. The data retrieved from such databases provides a comprehensive information on molecular weight, isoelectric pH, protein sequence and the sequence coverage of protein under study (Fig. 3.4e, f). To achieve high sequence coverage, maximum number of peaks should be analysed while performing MS-MS.

Various databases, such as UniProt (Swiss-Prot), NCBInr, PlasmoDB and PlantEST (Table 3.2), help in protein identification via search engines like Matrix science (Mascot server). Though MALDI is a reliable and effective technique for rapid identification of proteins, the databases play a significant role in data generation and analysis. For this reason, studies associated with proteins from remote sample sources with no record of their information in existing databases face challenges in their true identification. Nevertheless, MALDI-TOF offers a promising platform for rapid molecular identification with extensive applications in clinical diagnostics, biomarker detection and tissue imaging studies.

3.4 Liquid Chromatography-Mass Spectrometry

The technique LC-MS is a result of a successful alliance between two techniques, the liquid chromatography and mass spectrometry. Liquid chromatography separates a highly complex mixture

Table 3.2 Few commonly used software/tools for protein identification from MS-MS data

S. No.	Software	Features	URL site	References
1	Mascot	Protein identification using mass spectrometry data	http://www.matrixscience.com/	Palagi et al. (2006)
2	MS-Fit	Mining the sequence of the protein from MS data	prospector.ucsf.edu	Palagi et al. (2006)
3	SEQUEST	Interpretation of tandem mass spectra data for protein identification and amino acid sequence	http://fields.scripps.edu/sequest/	Palagi et al. (2006)
4	X! Tandem	Protein identification using tandem MS data	http://www.thegpm.org/tandem/index.html	Palagi et al. (2006)
5	Sequit!	De novo sequencing of protein using tandem mass spectrum	http://www.sequit.org/	Palagi et al. (2006)
7	PEAKS	De novo sequencing of raw MS-MS data, label-free quantification	http://www.bioinfor.com/	Pevtsov et al. (2006)

into different subsets based on their physical and chemical properties. The separated mixtures are then ionized and injected into the mass spectrometer where ions get further separated under the influence of a strong electromagnetic field before getting detected. The data obtained is in the form of a spectrum, which can either be analysed manually (if the sample contains very few proteins) or with the use of specialized software developed for analysing high-throughput data (Table 3.2). Despite the superior capabilities of this technique, its use in protein studies picked up pace only in the 1980s, after the advent of ionization techniques such as electrospray ionization. This soft ionization method was demonstrated to generate charged ions from peptides in solution thereby paving the way for the emergence of LC-MS as a robust technique in proteomics. Over the last two decades, LC-MS has played a critical role in proteomics and contributed in sequencing of the yeast proteome (Peng et al. 2003), human tissues and cell lines (Phanstiel et al. 2011; Munoz et al. 2011; Geiger et al. 2012; Moghaddas Gholami et al. 2013), to name a few.

3.4.1 Data Acquisition and Analysis

Before the complex protein sample is injected into the mass spectrometer, a number of processing steps, aimed at reducing the complexity of samples, are involved. The process of enzymatic digestion ensures that the complex proteins are broken down into less complex peptides. MS instruments measure the m/z values of various fragments by reducing these sequences into patterns of numbers. Therefore, understanding the number pattern becomes very important in identifying the right peptide sequence leading to the right protein. A clear idea of the chemistry behind fragmentation is needed to help marvel the ability of these highly sensitive instruments. When the peptide ions collide with the neutral gas molecules in the collision cell, the most common cleavage involves that of the peptide bond forming the backbone. This cleavage results in the formation of two ions – most commonly referred to as the “y ion” and the “b ion”. The “y ion” is positively charged and represents the C-terminal end of the peptide, while the “b ion” is negatively charged ion and represents the N-terminal of the peptide. It is to be noted that in addition to the peptide bond, the other bonds of less significance also undergo fragmentation, albeit less frequently.

Manual analysis of the spectra and data interpretation is easier with a definite pattern of observed peptide peaks. In cases of complex peptide peak profiles, the process of understanding the data and being able to decipher the sequence can be time consuming. Moreover, the ambiguous fragmentation of certain peptides due to the variable amino acid fragmentation tendency and presence of proline, which does not facilitate easy fragmentation, can pose additional challenges, thereby making manual data analysis from MS-MS spectra an uphill task.

Owing to challenges associated with manual analysis of tandem mass spectrometry data, a number of software have been developed, each with their own merits and demerits. While most commercially available software come with a huge price tag, software like Mascot allow easy analysis and interpretation of data from tandem mass spectrometry experiments.

We now take up an example of an LC-MS run for a protein sample from *Arabidopsis thaliana* to better understand, analyse and interpret an LC-MS data set. The foremost step in analysing data using Mascot involves filling up of a MS-MS ion search form on the Mascot website, enquiring all important information such as the enzyme used during digestion, database used for search, number of missed cleavages, taxonomy, fixed and variable modifications, MS-MS tolerance, peptide charge, etc. After initiating the search, the software generates a result file containing a Mascot histogram and a peptide summary report. The protein hits falling outside the green region of the histogram are considered significant. The generated peptide summary report allows the user to select the significant peptides, in addition to other valuable information such as molecular weight, total ion score and the number of peptides matched (Fig. 3.5a). The peptide view contains all information regarding the peptide sequence derived from the corresponding "y" and "b" ions (Fig. 3.5b, c).

In the current analysis, a total of three significant hits were obtained. Of these, one hit corresponded to trypsin used for digestion of proteins. The other two peaks corresponded to the following proteins: photosystem I reaction centre subunit IV A (chloroplastic) and photosystem I reaction centre subunit IV B (chloroplastic), with molecular masses 14,958 Da and 15,188 Da, respectively.

3.4.2 Advancements in LC-MS-Based Quantification of Proteins

The technological advancements in mass spectrometry have led to increased use of these new-age instruments in quantitative proteomics. Quantification of peptides can either be achieved through differential labels such as iTRAQ and ICAT (isotope-coded affinity tag) or SILAC or label-free quantification approaches. The use of labels allows relative quantification of peptides among different biological samples in a single run. The iTRAQ-labelling approach enables calculation of peptide abundance based on intensities of fragment ions reported in the obtained MS-MS spectra (Chloe et al. 2007). The ICAT-based approach (Gygi et al. 1999) and SILAC (Ong et al. 2002), on the other hand, result in the generation of pairs of peptides with mass differences characteristic to the label used. The shift in masses and similarities in elution profiles form the basis for computing peptide ratios between the two forms of labels thereby allowing determination of the relative abundance of peptides. Despite their widespread usage, a major limitation associated with these approaches is the prejudice towards high-intensity peptide signals for the selection of the precursor ion, which results in under-sampling of low-abundant proteins in the sample mixture (Mueller et al. 2008).

The past few years have seen an upsurge in the label-free quantification strategies, which rely on correlating the abundance of a protein or peptide in a sample with the corresponding MS signal (Simpson et al. 2009). Determination of ion intensity using extracted ion chromatograms (XIC) is a popular method for the protein quantification. This involves summation of the number and intensity of the selected precursor ions at a specific m/z range and peak areas as a measure of the relative abundance (Old et al. 2005). Another approach that is increasingly being used for quantification is spectral counting of fragment ion spectra for a particular peptide. This is a semi-quantitative approach used for low to moderately mass-resolved LC-MS data. The approach relies

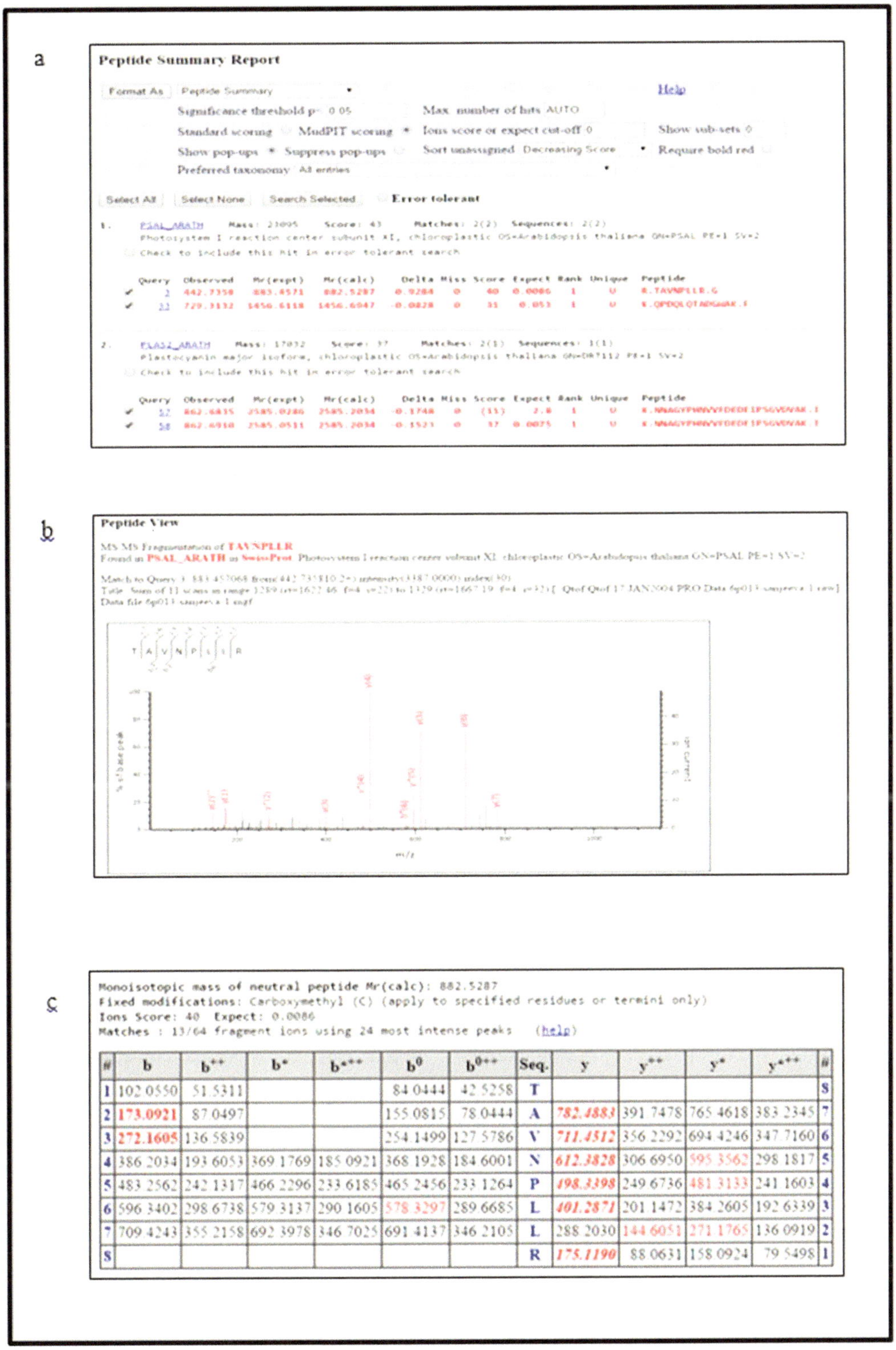

Fig. 3.5 MS-MS search results of an in-gel digested protein sample mixture using Mascot. The panels demonstrate (**a**) peptide summary report containing details of significant hits, (**b**) peptide view of a significant hit indicating the peptide sequence information and (**c**) the information regarding residual masses of amino acids deducing the peptide sequence

on the assumption that the frequency of a particular precursor ion getting selected in a large data set is proportional to the abundance of the peptide in a sample. The spectral counts from peptides are averaged and an abundance index of a protein is generated (Liu et al. 2004; Gao et al. 2003; Colinge et al. 2005; Ishihama et al. 2005). The label-free approaches mentioned above facilitate quantification of peptides without using expensive labels and performing additional sample processing steps. They are, hence, becoming the choice of most researchers despite their inherent limitations (Simpson et al. 2009). However, these approaches are still evolving, and are believed to improve greatly in the years to come (Mueller et al. 2008).

3.5 Protein Microarrays

Protein microarrays have been widely accepted as a high-throughput technique to achieve systemic understanding of protein-protein interactions, functional analysis of proteins and autoantibody screening in various systems (Mitchell 2002; MacBeath 2002). This technology essentially relies on proteins immobilized on glass slides, traditionally coated with PVDF (polyvinylidene fluoride), nitrocellulose or polystyrene. The approach has now evolved into incorporating soft lithography techniques to enhance surface chemistry for the immobilization of proteins (Hu et al. 2011). These protein arrays are subjected to a set of probe molecules, and are classified on the basis of the biological question to be answered. For instance, functional protein arrays are arrays where immobilized proteins are subjected to probing by query DNA, RNA, peptides, small molecules, glycans or protein molecules to observe their interaction with ligands on the chip (Phizicky et al. 2003; Hu et al. 2011). Analytical protein arrays are arrays on which ligands like allergens, aptamers, antibodies or antigens are printed to perform protein profiling or clinical diagnostics (Phizicky et al. 2003). Reverse-phase protein microarrays is another popular kind of biomarker validation platform, where a large number of clinical samples such as, biofluids or cell lysates are printed on the chip and probed with antibodies for target biomarkers for large-scale screening in clinical cohorts (Zha et al. 2004; Tibes et al. 2006).

Although there are several types of protein microarrays with varied applications for each, autoantibody screening from biofluids like serum and CSF has been one of the most popular applications for elucidating novel biomarkers in infectious diseases or cancers (Song et al. 2010; Anderson et al. 2011; Hu et al. 2012). Autoantibody production is a response of the immune system against certain aberrant self-proteins, also termed as tumour-associated antigens (TAAs) in case of cancer (Anderson et al. 2011). Biofluids can be screened for the presence of autoantibodies by high-throughput protein microarrays harbouring human peptides or whole proteins (Fig. 3.6a, b). If antibodies are produced against an aberrant protein, they would bind to the antigen and detected using Cy dye-labelled secondary antibodies displaying fluorescent signals. The spot intensity of the protein would indicate the strength of the immunogenic response against a particular protein, which enables relative quantification of the protein. The signal intensities are measured by scanning the chip using a microarray scanner at appropriate channels depending on the absorbance wavelength of the Cy dye employed in the assay (Fig. 3.6a). The data exported is usually in .gpr, .cel or .txt format, depending on the scanner used to generate the output file, which contains information regarding image acquisition and each protein spot (feature). As against tissue biopsy, a highly invasive diagnostic approach, the autoantibody profiling using serum is a minimal invasive approach aiding in early diagnosis of cancer.

DNA microarrays provided the foundation for data analysis strategies for protein microarrays. (Hu et al. 2011; An et al. 2014). Here, we will focus on the generic data analysis approach applicable across all platforms of protein microarrays. The workflow of protein microarrays for screening autoantibodies is very similar to other immunological assays like western blotting and ELISA. However, the staggering difference in microarray throughput as against to these

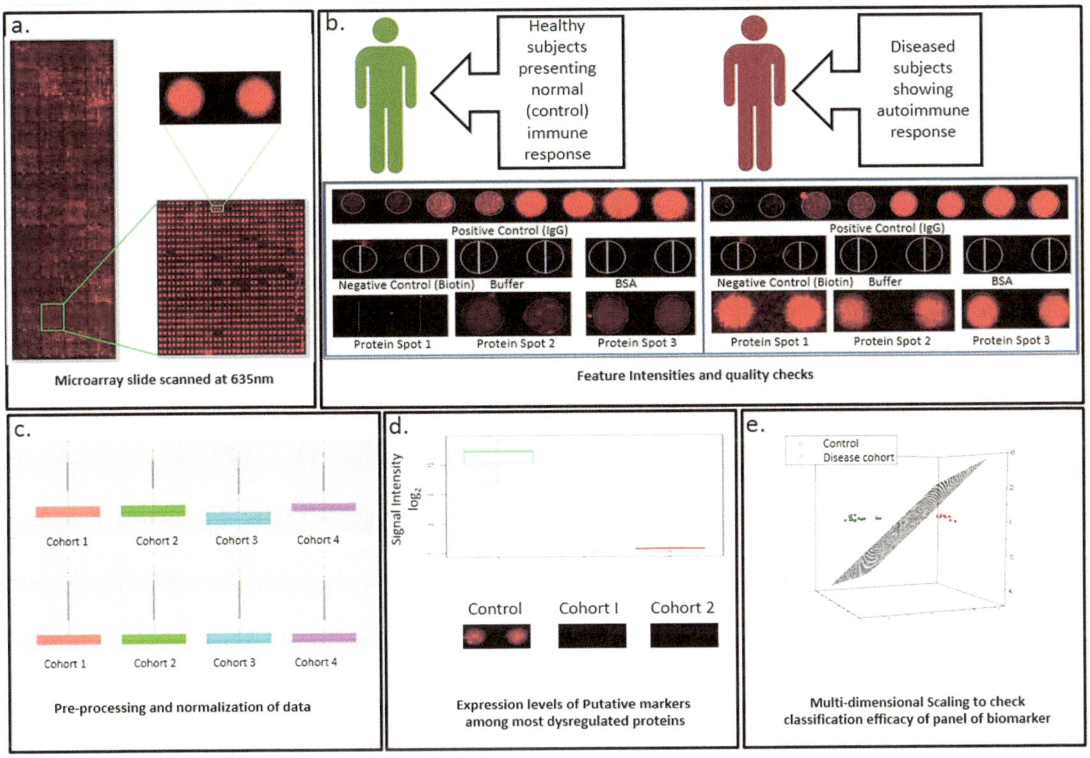

Fig. 3.6 An overview of protein microarray data analysis. The figures represent (**a**) a processed microarray slide scanned at 635 nm for Cy5. This results in illumination of features, signal intensities of which are extracted during data acquisition. The autoantibody profile of a healthy against diseased subject is shown in panel (**b**). Quality control features like positive controls and negative controls aid as landmarks in the normalization process. Differentially expressed proteins can visually be observed showing opposite trends in the two cohorts. The visual representation of unnormalized (*top panel*) and normalized data (*bottom panel*) is shown in panel (**c**). The differential expression of a putative protein marker emerging from the data, and its relative fold change across the cohorts is shown in (**d**) upper panel, whereas the bottom panel shows the visual spot intensities. Panel (**e**) represents the classification of subjects in two distinct cohorts based on the group of classifiers deduced using mathematical algorithms

traditional techniques, accompanied with other variables influencing the experimental outcome, makes protein microarray data analysis extremely challenging. Specialized software used for data acquisition from the microarray scanner, statistical models and robust computational support followed by systems biology approaches are the fundamentals of protein microarray data analysis (An et al. 2014). The primary statistical and computational elements of data analysis in a protein microarray experiment can be broadly divided into following stages.

3.5.1 Pre-processing of the Data: Background Correction and Normalization

The use of Cy dyes often results in false positives (noise), in addition to the true positives, due to non-specific binding. The true estimate of a spot intensity is obtained by subtracting the background intensity from the foreground, called background correction. In order to study the effects of different background correction methods, log-normalized foreground and background intensities are plotted for different samples without performing any correction at first. One of the methods (normexp + offset) from LIMMA (linear model for analysing differential expression) model is used to normally distribute the background intensities treating the foreground signal as an exponential distribution while stabilizing any resulting variance (Syed et al. 2015).

The underlying assumption of any microarray experiment is that the majority of proteins display the same expression levels across arrays. In order to study the biological differences, the technical variation that may arise due to dye bias, print-tip effects or day-to-day variations must first be optimized. It is therefore important that the data is normalized for an unbiased analysis (Fig. 3.6c). Some of the commonly used normalization strategies include quantile normalization, variance-stabilizing normalization, cyclic loess and robust-linear-model normalization (An et al. 2014). These are essentially mathematical algorithms aimed at distributing the variance arising in a set of arrays to normalize the signal intensities. Each of these strategies may be used under different set of conditions, depending on the nature of experiment in consultation with the biologists, clinicians and statisticians analysing the data. A comparative analysis of these normalization strategies has been described in one of the previous studies (Kingsmore 2006).

3.5.2 Differentially Expressed Proteins

Protein microarrays are generally used to comprehend the differential protein expression levels across any two cohorts (diseased vs. healthy) (Fig. 3.6d). Determination of differentially expressed proteins between two sets of samples involves statistical tests with a null hypothesis that no gene is differentially expressed. In order to screen for significant biomarkers that could differentiate these cohorts, a robust analysis of protein expression levels is required. The student's *t*-test (assuming normal distribution of data), rank product (non-parametric), Wilcoxon rank-sum test (assuming nonparametric, normal distribution approximation), significance analysis of microarrays (SAM), LIMMA and M-statistic are commonly employed for generating a list of differentially expressed proteins (An et al. 2014). However, since protein microarrays are highly dynamic, statisticians often choose one or a combination of these tools or alternatively develop complementary approaches to improve the stringency of data, especially if there are underlying assumptions of data distribution.

3.5.3 Shortlisting Differentially Regulated Proteins

Correction methods, like Benjamini-Hochberg, shortlist proteins based on their statistical significance providing a *p*-value cut-off. Another way to improve the stringency of shortlisting proteins is to employ a fold-change cut-off further to the corrected *p*-value cut-off. This is a manual way of examining if the data shortlisted qualifies the threshold cut-off, which may be of interest to a biologist. These shortlisted proteins could be studied further or subjected to algorithms like correspondence analysis (CA) (Syed et al. 2015), a model analogous to principal component analysis (PCA), to understand the degree of classification that can be achieved to segregate two cohorts. CA is a dimensionality reduction technique, which helps in narrowing down the long list of differentially expressed proteins. The list from correspondence analysis can be treated as set of markers whose values are associated with classes in a statistically significant manner rather than by mere chance. In order to select a panel of significant classifier proteins differentiating control samples from diseased samples, recursive feature elimination using models like support vector machine can be used (Syed et al. 2015). The efficacy of these models can be visually validated using multidimensional scaling plots (Fig. 3.6e) (Syed et al. 2015). Figure 3.7 describes the basic workflow of such an autoantibody screening experiment along with a general pipeline for protein microarray data analysis.

A biologist could use a systemic approach to understand the dysregulated pathways emerging

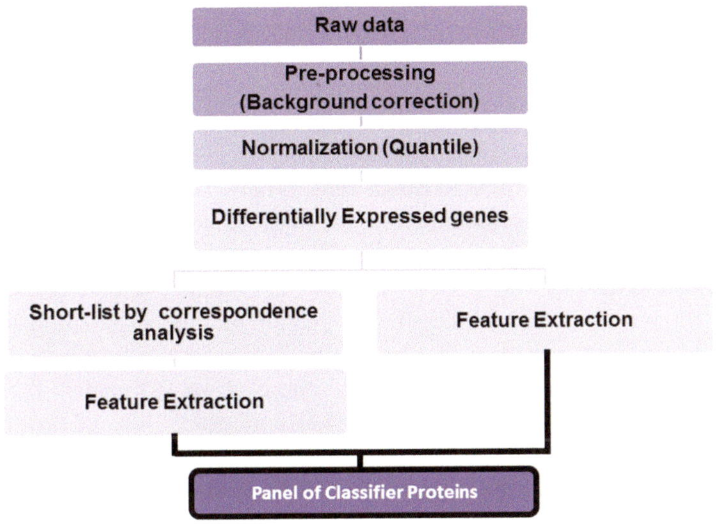

Fig. 3.7 Schematic representation of the protein microarray data analysis strategy

from the list of significantly dysregulated proteins. In addition to this, protein interaction and metabolic networks, gene ontology and gene set enrichment analysis can be performed to completely understand pathobiology of the disease under study (Syed et al. 2015). Classifier proteins with high fold-change values can be validated in clinical diagnostics. Thus protein microarray platform, with indispensable computational and statistical support for robust data analysis, is a powerful discovery tool for biomarker studies.

3.6 Surface Plasmon Resonance

Surface plasmon resonance (SPR) is an optical method to monitor changes in the refractive index of materials in the near vicinity of the metal surface. It is a phenomenon that occurs when polarized light, under the condition of total internal reflection, strikes an electrically conducting thin metal film at the interface between media of different refractive index: the glass of the sensor chip surface and the sample solution.

As the plane polarized light strikes through a high refractive index prism, the light becomes totally internally reflected and generates an evanescent wave that penetrate the thin metal film. At a certain angle of incidence, the incident light excites surface plasmons (electron charge density waves) on the metallic film. As a result, there is a characteristic absorption of energy via the evanescent wave field, and a drop in the intensity of reflected light at a specific angle known as the resonance angle. These surface plasmon waves are extremely sensitive to the refractive index of the solution within the effective penetration depth of the evanescent field. Interaction of biomolecules produces a change in the refractive index near the metal surface, leading to a shift in the resonance angle, which is monitored in real time by detecting changes in the intensity of the reflected light. The apparent rate constants for the association (K_a) and dissociation (K_d) can be analysed from the rate of change of the SPR signal.

SPR-based biosensors are now routinely used as an established platform for validating biomolecular interactions and performing concentration analysis (Pattnaik 2005; Helmerhorst et al. 2012; Berggård et al. 2007; Boozer et al. 2006; Shah et al. 2015). The technology allows analysis of these interactions in real time with high sensitivity and low sample requirement in a label-free environment. The method is not restricted to the usage of protein-protein interactions, but the generality extends to all kinds of molecules including protein-lipid, protein-RNA and protein-nanoparticle studies (Katsamba et al. 2002; Cedervall et al. 2007; Navratilova et al. 2006). Briefly, one of the interacting molecules (ligand) is bound on a sensor chip surface and the other interacting partner (analyte) is injected over the surface. The amount of analyte bound is continuously monitored as a function of time showing the progress of interaction. This plot of response against time is called sensorgram. The SPR response is proportional to the mass of analyte bound at the sensor surface. The analyte injection is followed by an increase in binding response which enables the determination of rate of complex formation (K_a). As the analyte injection is replaced by buffer flow, the rate of dissociation of the complex (K_d) can be monitored. The complex may not dissociate completely in many cases, wherein regeneration of the surface is performed with mild acidic or basic washing conditions.

3.6.1 SPR Data Processing and Analysis

In SPR, the data is collected continuously over time so that the kinetic parameters can be determined with accuracy and precision. SPR analysis of biomolecular interactions involves crucial experimental design and depends on several experimental factors such as optimum buffers, pH conditions, immobilization chemistry, ligand density, regeneration solutions, flow rate and temperature. In single-cycle kinetics, analytes with increasing concentrations are injected one after the other in a single cycle without the need to regenerate the surface between sample injections. However, in multi-cycle kinetics, different analyte concentrations are run as different cycles which may require regeneration of the surface after every individual cycle, depending on the dissociation pattern of the analyte molecule. Figure 3.8 demonstrates an example of protein interaction with a small drug molecule performed using both multi-cycle kinetics and single-cycle kinetics. The latter approach reduces the experimental time involved and seems aptly suitable for situations where optimum regeneration conditions cannot be achieved. It is suggested to immobilize low amount of ligand for kinetic analysis of macromolecular interactions to achieve surface saturation and avoid mass transport limitation effect and aggregation. Mass transport limitation occurs when the binding rate of analyte to ligand is faster than the diffusion rate of analyte to the ligand surface. Low surface immobilization and high flow rates can minimize this effect allowing better fit of models.

After the data is collected, several processing steps need to be performed before any quantitative information can be extracted. A number of software programs are available for processing SPR data, including BIAevaluation, Scrubber and CLAMP (Morton and Myszka 1998). The initial steps in data processing involve zeroing on the

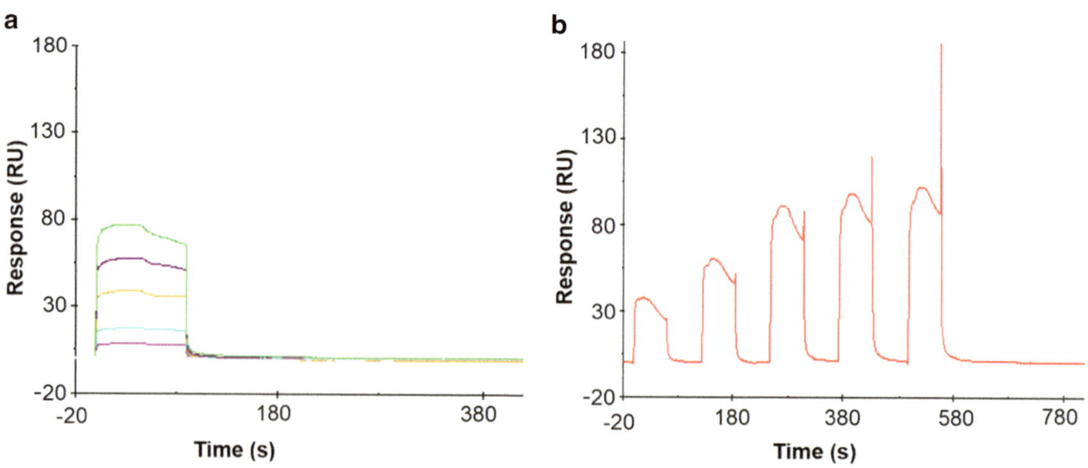

Fig. 3.8 Surface plasmon resonance analysis for a protein-drug interaction study. The figures illustrate protein interaction with a small drug molecule performed using (**a**) multi-cycle kinetics and (**b**) single-cycle kinetics

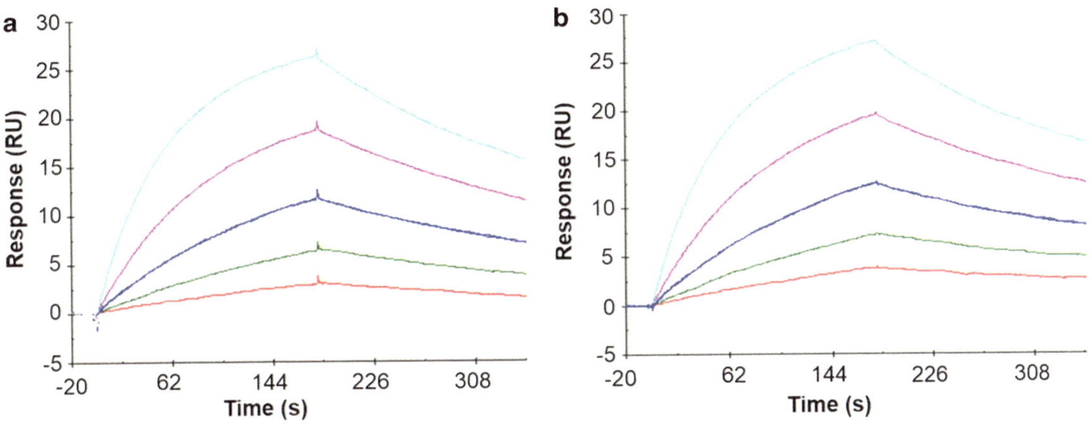

Fig. 3.9 An example illustrating blank subtraction for subtracting bulk effects and checking the specificity. A protein-protein kinetic sensorgram showing subtraction of an ideal baseline response (shown in *grey*) before sample injection from the obtained sample response of different concentrations (shown in different colours). The panel (a) shows the unsubtracted sensorgram, whereas panel (**b**) displays the blank-subtracted sensorgram

x-axis (time) and y-axis (response units). This aligns the beginning of injections with respect to each other and allows the responses observed from each flow channel to be compared with one other. The referencing steps help in minimizing artifacts, and also correcting for any bulk shift resulting from buffer mismatch in sample buffer and running buffer. In the first referencing step, the reference flow cell sensorgram is subtracted from the active flow cell to produce a sensorgram removing bulk shift contributions. In the second referencing step, as exemplified in Fig. 3.9 from a protein-protein interaction study, the effect of buffer injections is nullified by subtracting the baseline response before sample injection from the obtained sample response. These two referencing steps, known as double referencing, remove the systematic shifts and drifts in baseline, frequently observed in sensorgrams (Myszka 1999).

The data now becomes ready for fitting to appropriate models using a mathematical algorithm, and to further determine the kinetic parameters (K_a, K_d and K_D) and characterize the interaction. The chosen analyte concentration range should be wide enough to achieve surface saturation, the highest concentration being approximately five to ten times the K_D value. Purity of the ligand and analyte, immobilization heterogeneity, mass transfer effects, rebinding of analytes to ligand, buffer mismatch, inappropriate analyte range and complexity of biological systems can greatly influence the fitting of models. There are a number of kinetic models available that can fit the acquired data such as 1:1 Langmuir fit model, heterogeneity model, bivalent analyte model, conformational change model, etc. In general, the simplest model, the 1:1 Langmuir fit model (one ligand molecule interacts with one analyte molecule assuming that the interaction rate is not limited by mass transport) should be tried as the first attempt since most of the biological interactions occur in a 1:1 stoichiometry. There should be a valid justification for the use of other models, the results of which should be confirmed with other supporting experiments. Many factors need to be considered while deciding the correct fit model.

In a global fit, both association and dissociation data, for all analyte samples, are fit at the same time using a sum of squared residuals over every data point. The global approach is more efficient because there are fewer adjustable parameters, whereas a different R_{max} value is calculated for each curve in a local fit. After the fit is made, the curves need to be studied well to understand the accuracy of fit for the association and dissociation phase, and examine if the calculated R_{max} is within the expected range. R_{max} is the maximum feasible response that can be obtained for a specific interaction. The theoretical R_{max} for an interaction can be calculated based on the following formula: R_{max} = analyteMW/ligandMW

*R_L*Sm, where R_{max} is the maximum binding response (RU), R_L is the immobilization level, S_m is the stoichiometric ratio (number of binding sites per ligand) and MW is the molecular weight. In concept, the theoretical R_{max} should be the same for different analyte concentrations injected one after the other, if we do not consider the binding site loss due to harsh regeneration or incomplete regeneration. R_{eq} is the response obtained when binding between ligand and analyte reaches equilibrium. The equilibrium constant, K_D, can be calculated directly using steady-state or equilibrium analysis, where the rate of association equals the rate of dissociation. The response at equilibrium, R_{eq}, is measured over a given range of analyte concentrations, and the values are plotted against those analyte concentrations. The kinetic and equilibrium analyses performed on the same data set should ideally produce similar K_D values, which can reflect the confidence level of the data obtained.

Differences in sample buffer and running buffer result in bulk signal, which does not allow models to fit well to the data (Rich and Myszka 2010). Many times, when the analyte concentrations are not accurately known, the curve fittings using software can be misleading. One such example is demonstrated from a protein-small molecule interaction study (Fig. 3.10a). In such cases, a good fit of the analysed kinetic data can be confirmed by low chi^2 (less than 10 % of R_{max}), which is the average of squared differences between the measured data points and the corresponding fitted values. One chi^2 value, which gives a measure of the closeness of model fit, is determined for all curves fitted simultaneously. Residual plots, generated by some of the software, determine the accuracy of fitting even better than chi^2 values. The residual plot shows the difference in response between each data point for the experimental curves and the calculated curves. The shape and distribution of the residual plot give a better insight on the quality of fit to the chosen model. If there are systematic deviations between the experimental and fitted curves, the plot will indicate them by displacement from the zero line. Figure 3.10b demonstrates an ideal residual plot obtained from an antibody-protein interaction using BIAevaluation software. The guidelines are drawn in green to indicate the range of acceptability. Ideally, the noise level in the plot should be on the order of ±2 RU (Drescher et al. 2009).

3.7 Concluding Remarks

Omics platforms have emerged as powerful tools to help researchers look at biosystems with a global perspective. Innovations in technologies have broadened our existing knowledge, revealing the interplay of various biomolecules at the

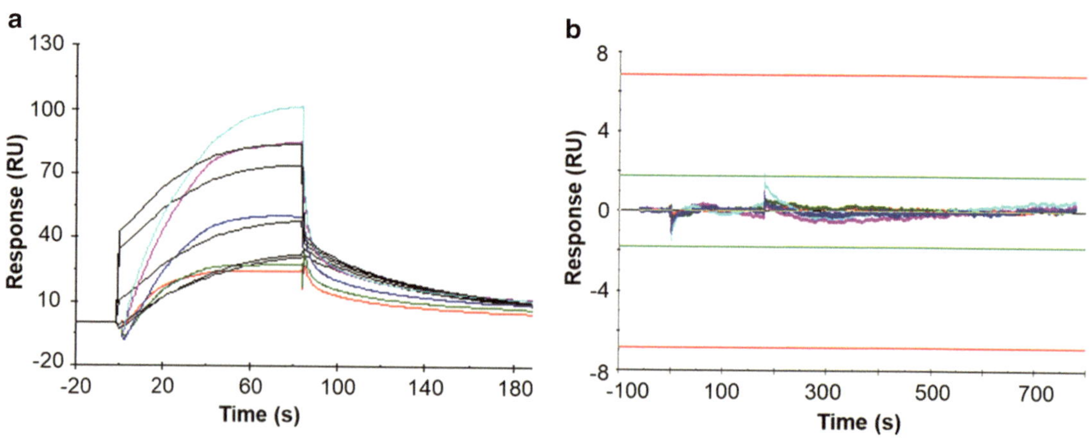

Fig. 3.10 An example illustrating a poor curve fitting and an ideal residual plot. A protein-small molecule interaction study resulting in (**a**) a poor curve fitting. An ideal (**b**) residual plot from an antibody-protein interaction showing differences in response between experimental and calculated curves, demonstrating the quality of fit to the chosen model

systemic level. However, the quantity of data assimilated through these technologies employs new challenges on data processing and analysis. High-throughput techniques aiming at decoding the complexity of biosystems have led to a surge of data, albeit with many downstream hurdles in the form of data storage and data analysis. In the last few decades, huge efforts have been devoted towards creation of database repositories for different data sets where researchers are encouraged to share their data associated with scientific publications. This has enabled researchers around the world to reproduce and validate the studies, as well as analyse the data in innovative ways using different methodologies. With multi-omics technologies routinely being used for various studies, it is important to appraise the challenges of data analysis associated with these sophisticated platforms. Data exploitation requires vital support from sophisticated software and explorative tools, employing statistical methods and visualization aids, to analyse heterogeneous data sets.

Lately, there have been significant advancements in proteomic techniques offering greater sensitivity and rapidity, complementing the traditional methods. Data processing and analysis in proteomics are certainly a complex multistep process. Common proteomic techniques like gel-based approaches, mass spectrometry, protein microarrays and label-free technologies find overlapping applications in multi-omics disciplines. As discussed above, these techniques are often employed for proteome profiling, identification of post-translational modifications, comparative expression analysis of proteins and studying molecular interactions. Accurate and reliable data processing and analysis are the fundamentals of these proteomic approaches to generate factual biological insights. Hence, data processing and analysis of heterogeneous data types is presently an active field of research where biologists and biostatisticians are persistently working together towards improving data utilization in research and discovery.

References

An LTT, Pursiheimo A, Moulder R et al (2014) Statistical analysis of protein microarray data: a case study in type 1 diabetes research. J Proteomics Bioinform S12:003. doi:10.4172/jpb.S12-003

Anderson KS, Sibani S, Wallstrom G et al (2011) Protein microarray signature of autoantibody biomarkers for the early detection of breast cancer. J Proteome Res 10:85–96. doi:10.1021/pr100686b

Beckett P (2012) The basics of 2D DIGE. In: Cramer R, Westermeier R (eds) Difference Gel Electrophoresis (DIGE): methods and protocols. Springer Protocols, New York, pp 9–19

Baggerman G, Vierstraete E, De Loof A, Schoofs L (2005) Gel-based versus gel-free proteomics: a review. Comb Chem High Throughput Screen 8:669–677

Berggård T, Linse S, James P (2007) Methods for the detection and analysis of protein–protein interactions. Proteomics 7:2833–2842. doi:10.1002/pmic.200700131

Boozer C, Kim G, Cong S et al (2006) Looking towards label-free biomolecular interaction analysis in a high-throughput format: a review of new surface plasmon resonance technologies. Curr Opin Biotechnol 17:400–405. doi:10.1016/j.copbio.2006.06.012

Cedervall T, Lynch I, Lindman S et al (2007) Understanding the nanoparticle-protein corona using methods to quantify exchange rates and affinities of proteins for nanoparticles. Proc Natl Acad Sci USA 104:2050–2055. doi:10.1073/pnas.0608582104

Chandramouli K, Qian P-Y (2009) Proteomics: challenges, techniques and possibilities to overcome biological sample complexity. Hum Genomics Proteomics HGP. doi:10.4061/2009/239204

Choe L, Ascenzo MD', Relkin NR et al (2007) 8-plex quantitation of changes in cerebrospinal fluid protein expression in subjects undergoing intravenous immunoglobulin treatment for Alzheimer's disease. Proteomics 7:3651–3660. doi:10.1002/pmic.200700316

Colinge J, Chiappe D, Lagache S et al (2005) Differential proteomics via probabilistic peptide identification scores. Anal Chem 77:596–606. doi:10.1021/ac0488513

Croxatto A, Prod'hom G, Greub G (2012) Applications of MALDI-TOF mass spectrometry in clinical diagnostic microbiology. FEMS Microbiol Rev 36:380–407. doi:10.1111/j.1574-6976.2011.00298.x

de Hoffmann E, Stroobant V (2007) Mass spectrometry: principles and applications. Wiley, New York

Drescher DG, Ramakrishnan NA, Drescher MJ (2009) Surface plasmon resonance (SPR) analysis of binding interactions of proteins in inner-ear sensory epithelia. Methods Mol Biol Clifton NJ 493:323–343. doi:10.1007/978-1-59745-523-7_20

Gao J, Opiteck GJ, Friedrichs MS et al (2003) Changes in the protein expression of yeast as a function of carbon source. J Proteome Res 2:643–649

GE Healthcare 2-D Electrophoresis Principles and Methods (2004) (online) http://www.med.unc.edu/pharm/sondeklab/Lab%20Resources/protein_purification_handbooks/2D%20electrophoresis.pdf

GE Healthcare DeCyder 2D Software, Version 6.5 User Manual (online) https://www.gelifesciences.com/gehcls_images/GELS/Related%20Content/Files/1314750913712/litdoc28401006_20131103235809.pdf

Gehlenborg N, O'Donoghue SI, Baliga NS et al (1998) Visualization of omics data for systems biology. Nat Methods 7:S56–S68. doi:10.1038/nmeth.1436

Geiger T, Wehner A, Schaab C et al (2012) Comparative proteomic analysis of eleven common cell lines reveals ubiquitous but varying expression of most proteins. Mol Cell Proteomics 11:M111.014050. doi:10.1074/mcp.M111.014050

Gomez-Cabrero D, Abugessaisa I, Maier D et al (2014) Data integration in the era of omics: current and future challenges. BMC Syst Biol 8:I1. doi:10.1186/1752-0509-8-S2-I1

Gygi SP, Rist B, Gerber SA et al (1999) Quantitative analysis of complex protein mixtures using isotope-coded affinity tags. Nat Biotechnol 17:994–999. doi:10.1038/13690

Hao C, Ma X, Fang S et al (1998) Positive- and negative-ion matrix-assisted laser desorption/ionization mass spectrometry of saccharides. Rapid Commun Mass Spectrom 12:345–348. doi:10.1002/(SICI)1097-0231(19980415)12:7<345::AID-RCM165>3.0.CO;2-B

Helmerhorst E, Chandler DJ, Nussio M, Mamotte CD (2012) Real-time and label-free bio-sensing of molecular interactions by surface plasmon resonance: a laboratory medicine perspective. Clin Biochem Rev Aust Assoc Clin Biochem 33:161–173

Hu S, Xie Z, Qian J et al (2011) Functional protein microarray technology. Wiley Interdiscip Rev Syst Biol Med 3:255–268. doi:10.1002/wsbm.118

Hu C-J, Song G, Huang W et al (2012) Identification of new autoantigens for primary biliary cirrhosis using human proteome microarrays. Mol Cell Proteomics 11:669–680. doi:10.1074/mcp.M111.015529

Ishihama Y, Oda Y, Tabata T et al (2005) Exponentially modified protein abundance index (emPAI) for estimation of absolute protein amount in proteomics by the number of sequenced peptides per protein. Mol Cell Proteomics 4:1265–1272. doi:10.1074/mcp.M500061-MCP200

Christian A Jackson WJS (1997) Application of mass spectrometry to the characterization of polymers. Curr Opin Solid State Mater Sci 661–667. doi: 10.1016/S1359-0286(97)80006-X

Katsamba PS, Park S, Laird-Offringa IA (2002) Kinetic studies of RNA-protein interactions using surface plasmon resonance. Methods San Diego Calif 26:95–104. doi:10.1016/S1046-2023(02)00012-9

Kempka M, Sjödahl J, Björk A, Roeraade J (2004) Improved method for peak picking in matrix-assisted laser desorption/ionization time-of-flight mass spectrometry. Rapid Commun Mass Spectrom 18:1208–1212. doi:10.1002/rcm.1467

Kingsmore SF (2006) Multiplexed protein measurement: technologies and applications of protein and antibody arrays. Nat Rev Drug Discov 5:310–320. doi:10.1038/nrd2006

Liu Q, Krishnapuram B, Pratapa P, et al (2003) Identification of differentially expressed proteins using MALDI-TOF mass spectra. In: Conference record of the thirty-seventh asilomar conference on signals, systems and computers, 2004. pp 1323–1327 vol.2

Liu H, Sadygov RG, Yates JR (2004) A model for random sampling and estimation of relative protein abundance in shotgun proteomics. Anal Chem 76:4193–4201. doi:10.1021/ac0498563

MacBeath G (2002) Protein microarrays and proteomics. Nat Genet 32(Suppl):526–532. doi:10.1038/ng1037

Marvin LF, Roberts MA, Fay LB (2003) Matrix-assisted laser desorption/ionization time-of-flight mass spectrometry in clinical chemistry. Clin Chim Acta Int J Clin Chem 337:11–21

Mitchell P (2002) A perspective on protein microarrays. Nat Biotechnol 20:225–229. doi:10.1038/nbt0302-225

Moghaddas Gholami A, Hahne H, Wu Z et al (2013) Global proteome analysis of the NCI-60 cell line panel. Cell Rep 4:609–620. doi:10.1016/j.celrep.2013.07.018

Morton TA, Myszka DG (1998) Kinetic analysis of macromolecular interactions using surface plasmon resonance biosensors. Methods Enzymol 295:268–294

Mueller LN, Brusniak M-Y, Mani DR, Aebersold R (2008) An assessment of software solutions for the analysis of mass spectrometry based quantitative proteomics data. J Proteome Res 7:51–61. doi:10.1021/pr700758r

Munoz J, Low TY, Kok YJ et al (2011) The quantitative proteomes of human-induced pluripotent stem cells and embryonic stem cells. Mol Syst Biol 7:550. doi:10.1038/msb.2011.84

Myszka DG (1999) Improving biosensor analysis. J Mol Recognit 12:279–284. doi:10.1002/(SICI)1099-1352(199909/10)12:5<279::AID-JMR473>3.0.CO;2-3

Navratilova I, Dioszegi M, Myszka DG (2006) Analyzing ligand and small molecule binding activity of solubilized GPCRs using biosensor technology. Anal Biochem 355:132–139. doi:10.1016/j.ab.2006.04.021

Old WM, Meyer-Arendt K, Aveline-Wolf L et al (2005) Comparison of label-free methods for quantifying human proteins by shotgun proteomics. Mol Cell Proteomics 4:1487–1502. doi:10.1074/mcp.M500084-MCP200

Ong S-E, Blagoev B, Kratchmarova I et al (2002) Stable isotope labeling by amino acids in cell culture, SILAC, as a simple and accurate approach to expression proteomics. Mol Cell Proteomics 1:376–386

Palagi PM, Hernandez P, Walther D, Appel RD (2006) Proteome informatics I: bioinformatics tools for processing experimental data. Proteomics 6:5435–5444. doi:10.1002/pmic.200600273

Palzkill T (2002) Proteomics. Kluwer Academic Publishers, New York

Pattnaik P (2005) Surface plasmon resonance: applications in understanding receptor-ligand interaction. Appl Biochem Biotechnol 126:79–92

Peng J, Elias JE, Thoreen CC et al (2003) Evaluation of multidimensional chromatography coupled with tandem mass spectrometry (LC/LC-MS/MS) for large-scale protein analysis: the yeast proteome. J Proteome Res 2:43–50

Pevtsov S, Fedulova I, Mirzaei H et al (2006) Performance evaluation of existing de novo sequencing algorithms. J Proteome Res 5:3018–3028. doi:10.1021/pr060222h

Phanstiel DH, Brumbaugh J, Wenger CD et al (2011) Proteomic and phosphoproteomic comparison of human ES and iPS cells. Nat Methods 8:821–827. doi:10.1038/nmeth.1699

Phizicky E, Bastiaens PIH, Zhu H et al (2003) Protein analysis on a proteomic scale. Nature 422:208–215. doi:10.1038/nature01512

Rich RL, Myszka DG (2010) Grading the commercial optical biosensor literature-class of 2008: "The Mighty Binders". J Mol Recognit JMR 23:1–64. doi:10.1002/jmr.1004

Shah VG, Ray S, Karlsson R, Srivastava S (2015) Calibration-free concentration analysis of protein biomarkers in human serum using surface plasmon resonance. Talanta 144:801–808. doi:10.1016/j.talanta.2015.06.074

Simpson KL, Whetton AD, Dive C (2009) Quantitative mass spectrometry-based techniques for clinical use: biomarker identification and quantification. J Chromatogr B Anal Technol Biomed Life Sci 877:1240–1249. doi:10.1016/j.jchromb.2008.11.023

Song Q, Liu G, Hu S et al (2010) Novel autoimmune hepatitis-specific autoantigens identified using protein microarray technology. J Proteome Res 9:30–39. doi:10.1021/pr900131e

Syed P, Gupta S, Choudhary S, et al (2015) Autoantibody profiling of gliomas to identify biomarkers using human proteome arrays. Sci Rep (in press)

Tibes R, Qiu Y, Lu Y et al (2006) Reverse phase protein array: validation of a novel proteomic technology and utility for analysis of primary leukemia specimens and hematopoietic stem cells. Mol Cancer Ther 5:2512–2521. doi:10.1158/1535-7163.MCT-06-0334

Unlü M, Morgan ME, Minden JS (1997) Difference gel electrophoresis: a single gel method for detecting changes in protein extracts. Electrophoresis 18(11):2071–2077

Webster J, Oxley D (2012) Protein identification by MALDI-TOF mass spectrometry. In: Zanders ED (ed) Chemical genomics and proteomics. Humana Press, Totowa, pp 227–240

Wilkins MR (2008) Proteome research: concepts, technology and application. Springer, New York

Zha H, Raffeld M, Charboneau L et al (2004) Similarities of prosurvival signals in Bcl-2-positive and Bcl-2-negative follicular lymphomas identified by reverse phase protein microarray. Lab Invest 84:235–244. doi:10.1038/labinvest.3700051

Basics of Mass Spectrometry and Its Applications in Biomarker Discovery

4

Panga Jaipal Reddy, Kishore Gollapalli, Saicharan Ghantasala, Tumpa Das, Sandip Kumar Patel, Venkatesh Chanukuppa, Sanjeeva Srivastava, and Srikanth Rapole

Abstract

Mass spectrometry (MS) is the method of choice for both qualitative and quantitative high-throughput proteome analysis. In the early years, mass spectrometry was used only for small molecule analysis. However, advances in ionization sources, mass analyzers, and mass detectors made MS the central force in proteomics technologies. Starting from its use in peptide mass fingerprinting (PMF), MS has evolved greatly over the last two decades and now finds use in shotgun proteomics where thousands of proteins can be quantified at once. Currently, MS finds use in targeted proteome analysis and is widely used for biomarker discovery in cancer, diabetes, cardiovascular diseases, and infectious diseases worldwide to diagnose the diseases at the early stage or to unravel the mechanism of pathogenesis. The applications of MS have not been limited to proteomics and have moved to metabolomics, lipidomics, tissue imaging, in understanding posttranslational modifications (PTMs), etc. This chapter provides details of mass spectrometry and its applications in biomarker discovery.

Keywords

Mass spectrometry • Cancer • Biomarker • PTM • MALDI • ESI

4.1 Introduction to Mass Spectrometry

Genome sequencing during the early 1980s and 1990s offered a ray of hope to improve our understanding about life. The human genome project is touted to be one of the largest scientific endeavors in the history of mankind and was aimed at deciphering the complete information to monitor

P.J. Reddy • K. Gollapalli • S. Ghantasala • T. Das • S.K. Patel • S. Srivastava
Department of Biosciences and Bioengineering, Indian Institute of Technology Bombay, Powai, Mumbai 400076, Maharashtra, India

V. Chanukuppa • S. Rapole (✉)
Proteomics Lab, National Centre for Cell Science, Ganeshkhind, Pune 411007, Maharashtra, India
e-mail: rsrikanth@nccs.res.in

S. Srivastava (ed.), *Biomarker Discovery in the Developing World: Dissecting the Pipeline for Meeting the Challenges*, DOI 10.1007/978-81-322-2837-0_4

disease diagnosis and prognosis (Venter et al. 2001; Lander et al. 2001). Thus far, genomes for more than 500 organisms have been sequenced. Despite the enormous data that genome sequencing could generate, there was little information about proteins, the end products of gene expression with roles in biological function, and maintaining homeostasis. It was now, that the use of different proteomic platforms and techniques to study proteins gained importance. Proteomic studies aimed at offering better insights into understanding the biological systems, answering the ambiguity created by the data from whole genome sequencing of different organisms, and strengthening the findings at the genomic as well as the transcriptomic level. Two-dimensional gel electrophoresis (2-DE) and differential in gel electrophoresis (DIGE) were the two gel-based approaches that were largely used in the 1980s and 1990s for most proteomic studies. However, the inherent limitations of these techniques and the relatively less data that these would generate meant that new, efficient, and fast techniques be employed in a bid to revolutionize the field of proteomics. It was during the early years of 1990s that mass spectrometry, a technique largely applied for small molecules, came to the rescue of the proteomics community by helping in large-scale protein identification (Aebersold and Mann 2003)(Table 4.1).

The technique of mass spectrometry (MS) finds use in protein identification and characterization. A typical mass spectrometer has three parts, namely, an ionization source responsible for ionization of target molecules, a mass analyzer to resolve ions based on mass to charge (m/z) ratio, and a detector to capture the ions and generate MS spectrum. In addition to these, the pre-fractionation techniques employed to reduce complexity of the samples and the various bioinformatics tools used also play a significant role in the outcome of an experiment. Over the last two decades, there have been rapid advancements in mass spectrometry with significant improvements in various components such as ionization source, mass analyzers, and detectors along with bioinformatics and pre-fractionation methodologies employed, thereby collectively making the MS-based proteomics techniques more robust.

4.1.1 Ionization Source

Mass spectrometry is applicable only to ionized molecules and due to this reason, an ionization source becomes almost indispensable for a mass spectrometer. In the early years of mass spectrometry, only hard ionization methods were available to create ions, and hence, the technique had limited or no use for studying biomolecules like proteins or peptides. However, in the early 1990s, the discovery of two soft ionization methods, i.e., matrix-assisted laser desorption and ionization (MALDI) and electrospray ionization (ESI), entirely changed the world of mass spectrometry (Hillenkamp and Karas 1990; Fenn et al. 1989). MALDI employs the use of laser to ionize the molecules, though the principle behind it is not very well understood. Laser bombardment leads to excitation of the matrix molecules (aromatic acids) causing a transfer of energy to the neighboring acidic analyte molecules (TFA or formic acid is used to maintain the acidic pH) resulting in ionization. The rapid bombardment of laser generates heat which desorbs the matrix and [analyte + H^+] in the gaseous phase. The generated ions travel from source plate to the detector in vacuum due to the difference in electrical potential at two ends. The ions generated in MALDI are mostly singly charged and primarily used for protein identification and molecular weight determination (Laiko et al. 2000).

Electrospray ionization (ESI) method is used for ionization of samples in solution and is mostly hyphenated to liquid chromatography (HPLC or nano-LC). The beauty of this technique is that it generates multiple-charged species effective in extending the mass range of analyzer. The process of ESI involves three steps, namely: (1) Dispersion of droplets: The liquid sample coming from Taylor cone maintained at high voltage (2–6 kV) and surrounded by nebulizing gas (dry nitrogen gas) results in formation of smaller-sized charged droplets. (2) Solvent evaporation: The nebulizing gas (dry nitrogen gas) and the

Table 4.1 Application of mass spectrometry in biomarker discovery

Disease (reference)	MS platform	Sample source	Proteins identified
Cancer			
Endometrial cancer (DeSouza et al. 2005)	Nano-LC system with API QSTAR Pulsar QqTOF (Applied Biosystems)	Endometrial tissue	Chaperonin 10, pyruvate kinase M1 or M2 isozyme, calgizzarin, heterogeneous nuclear ribonucleoprotein D0, macrophage migratory inhibitory factor, and polymeric immunoglobulin receptor precursor; those that were under expressed are alpha-1-antitrypsin precursor, creatine kinase B, and immunoglobulin receptor transgelin
Ovarian cancer (Boylan et al. 2010)	LC system with a QSTAR Pulsar MS TOF-MS (Applied Biosystems)	Serum, OC $n=60$, $H^*=60$	Extracellular matrix protein-1, leucine-rich alpha-2 glycoprotein-1, lipopolysaccharide binding protein-1, and proteoglycan-4
Pancreatic cancer (Tonack et al. 2013)	Nano-LC-ESI-MS/MS, on a QSTAR Pulsar i Hybrid Mass Spectrometer (AB Sciex)	Serum $n=59$ and pancreatic juice $n=25$	Complement C5, inter-α-trypsin inhibitor heavy chain H3, α1-β glycoprotein, and polymeric immunoglobulin receptor
Prostate cancer (Rehman et al. 2012)	ESI-qQTOF-MS/MS (Applied Biosystems)	Serum $n=20$, tissue $n=56$ and cell lines $n=4$	Elongation factor 1 alpha 1 (eEF1A1)
Lung cancer (Keshamouni et al. 2006)	MALDI-TOF/TOF (Applied Biosystems)	EMT of epithelial cells	Tropomyosins filamin A, B, and C, integrin-â β1, heat shock protein 27, transglutaminase2, cofilin, 14-3-3 zeta, ezrin-radixin-moesin
Breast cancer (Leong et al. 2006)	ESI-QTOF-MS/MS (Applied Biosystems); LC interface with QqTOF	Cell lines, mitochondria, and ER	PPIB, AHNAK, and SLC1A5
Gastric cancer Mu et al. (2014)	Nano-LC coupled to QStarXL mass spectrometer (Applied Biosystems)	Cell lines ($n=2$)	Sorcin isoform b, multidrug resistance protein 1, heat shock protein beta-1, 14-3-3 protein gamma, major vault protein, DNA replication licensing factor, and macrophage migration inhibitory factor
Cervical cancer (Raemdonck et al. 2014)	HPLC-MALDI-TOF/TOF	Cervicovaginal fluid (n=29), precancerous = 6, $H^*=6$	Alpha-actinin-4 and pyruvate kinase isozyme
Oral cancer (Hu et al. 2008)	Nano-LC system with QqTOF mass spectrometer (Applied Biosystems)	Saliva (OC OSCC = 64 and $H^*=64$)	M2BP, MRP14, CD59, catalase, and profilin

(continued)

Table 4.1 (continued)

Disease (reference)	MS platform	Sample source	Proteins identified
Head-and-neck cancer (Ralhan et al. 2008)	QSTAR Pulsar i mass spectrometer (Applied Biosystems)	Tissue CS =9, *H**=5	Stratifin, YWHAZ, three calcium-binding proteins of the S100 family, S100-A2, S100-A7, and S100-A11, prothymosin, L-lactate dehydrogenase A chain, glutathione S-transferase Pi, APC-binding protein EB1, and fascin. Peroxiredoxin2, carbonic anhydrase I, flavin reductase, histone H3, and polybromo-1D
Diabetes mellitus			
Diabetic retinopathy (Torok et al. 2013)	Nano-HPLC-coupled ESI-MS/MS mass spectrometry	DR tear *n*=119, H tear = 65 55; 165 eyes	
Type 2 diabetes (Hwang et al. 2010)	ESI-QTOF (Applied Biosystems) HPLC – ESI-MS/MS	Skeletal muscle *n*=8	Desmin and alpha-actinin-2
Diabetes mellitus (Busto et al. 2008)	ESI-QTOF (Applied Biosystems)	Blood sample *n*=11	Glycated hemoglobin
(Type 2 diabetes mellitus Kaur et al. 2012)	ESI-QTOF (ABSCIEX) coupled with nanoacquity – UPLC system	Serum n D-2 = 106,n T2D = 106, *H**=76	Apolipoprotein-AI, vitamin D-binding protein, fibronectin, afamin, transthyretin, and β-galactosidase
(Type 2 diabetes mellitus Rao et al. 2009)	LTQ ion trap mass spectrometer (Thermo)	DM saliva *n*=10 and *H**=10; DM=10, IGT = 10, IGT + IFG = 10, and *H**=10	Alpha-1-antitrypsin, cystatin C, alpha-2-macroglobulin, and transthyretin
Cardiovascular disease			
Cardiovascular disease (Magiera et al. 2013)	Q Trap triple quadrupole (Applied Biosystems)	Urine, n cardiac failure = 10, *H**=10	Ketoglutaric acid (-KG), l-carnitine (l-CAR), and acetyl-l-carnitine (acetyl-l-CAR)
Cardiovascular disease (Keshishian et al. 2009)	Q Trap hybrid triple quadrupole/linear ion mass spectrometer (Applied Biosystems)	Blood *n*=6	CRP, MRP14, MPO, cTnT, cTnI, and NTproBNP
Cardiovascular disease (Yan et al. 2014)	Q Exactive quadrupole orbitrap mass spectrometer (Thermo Fisher Scientific)	Serum (CHD n=2009;130, *H**=130)	SAA (upregulated), apoC-I (downregulated), and HDL group protein
Cardiovascular disease (Grant et al. 2009)	MALDI-TOF-TOF (Applied Biosystems)	Serum and tissue (CHD n=6, *H**=6) aged = 6, young = 6	Calnexin, prohibitin, and VDAC1
Cardiovascular disease (Zheng et al. 2014)	LC-MS connected to LTQ-Orbitrap XL mass spectrometer (Thermo)	Saliva	Alpha-2-HS-glycoprotein precursor, isoform I of fibrinogen alpha chain precursor, and tubulin alpha-4A chain

(continued)

Table 4.1 (continued)

Disease (reference)	MS platform	Sample source	Proteins identified
Infectious diseases			
Chikungunya (Puttamallesh et al. 2013)	LTQ-Orbitrap (Thermo Scientific)	Serum- CHIKV infected = 5 and H^*=4	Clusterin, apolipoproteins, and S100A
Dengue (Fragnoud et al. 2012)	Ion trap esquire HCT ultra (Bruker Daltonics GmbH)	Plasma (DF = 5 DHF=5) SD=5, DF=5	Leucine-rich glycoprotein 1, vitamin D-binding-protein, and ferritin
Falciparum malaria (Huang et al. 2012)	QTOF 6520 (Agilent Technologies)	Saliva (FM = 17 and H^*=17)	PFL0480w, PF08_0054, and PFI0875w
Malaria (Mu et al. 2014)	LTQ XL™ mass spectrometer	Serum (KM=9, VM=6, and FM=10)	Adhesion molecule-4 and C-reactive protein
Tuberculosis (Liu et al. 2013b)	MALDI-TOF MS) (Kratos Analytical Co, UK)	Serum (YB=211 and H^* =180) TB=180, control = 211 (H^*=91, LC=40, pneumonia = 40, COPD = 40)	Fibrinogen, alpha polypeptide isoform alpha-E preproprotein

*Healthy control

temperature applied in the source causes evaporation of the solvent leading to a decrease in the size of droplets and increase in the charge density on the droplet. (3) Ionization: At critical point, the droplet becomes kinetically and energetically unstable causing the charged species to be ejected into gaseous phase and the potential difference makes ions enter into analyzer (Felitsyn et al. 2002; Iribarne and Thomson 1976).

4.1.2 Mass Analyzer

Mass analyzer forms the most important component of a mass spectrometer and is the part where the ions get resolved in vacuum. The commonly used mass analyzers include time of flight (TOF), quadrupole (Q), ion trap, orbitrap, and ion cyclotron resonance (ICR). Each mass analyzer is unique in terms of its principle, mass resolution, properties, sensitivity, dynamic range, etc. (Han et al. 2008). In the initial years of mass spectrometry, a typical mass spectrometer consisted of a single mass analyzer and would find use only for protein identification based on peptide mass fingerprinting (PMF). However, with advancements in technology commercial mass spectrometers with two or more mass analyzers in a series separated by collision cell commonly termed "tandem," mass spectrometers became increasingly available. Tandem mass spectrometry improves the selectivity and structural elucidation by providing amino acid sequence information. Q-TOF, triple quadruple (QqQ), TOF/TOF, LTQ-Orbitrap, Q-iontrap, and LTQ-FTICR are the commonly used mass analyzers finding use for different proteomics applications such as shotgun proteomics (protein coverage), quantification (label-free/iTRAQ/TMT/SILAC), and targeted proteomics (SRM/PRM/MRM) (Domon and Aebersold 2006) (Fig. 4.1). In the shotgun proteomic approach, the first mass analyzer operates in "radio frequency" mode where it allows elution of selected peptide from liquid chromatography column. In collision cell, the selected peptide also called as "precursor" is fragmented into product ions called as "daughter ions." Here, the cationic peptides in gas phase collide with inert gas (argon/helium/nitrogen) atoms leading to fragmentation by the process of collision-induced dissociation (CID) of precursor. The lowest energy peptide bond of the precursor is targeted to generate y and b ions (sometimes breaking of C–C and C–N bonds generates a, z, c, and x ions in addition to the "satellite" fragment ions such as y-NH_3, b-H_2O). In addition to CID described above, electron transfer dissociation (ETD) or electron capture dissociation (ECD) methods are also used widely for special applications like characterization of phosphorylation, glycosyl-

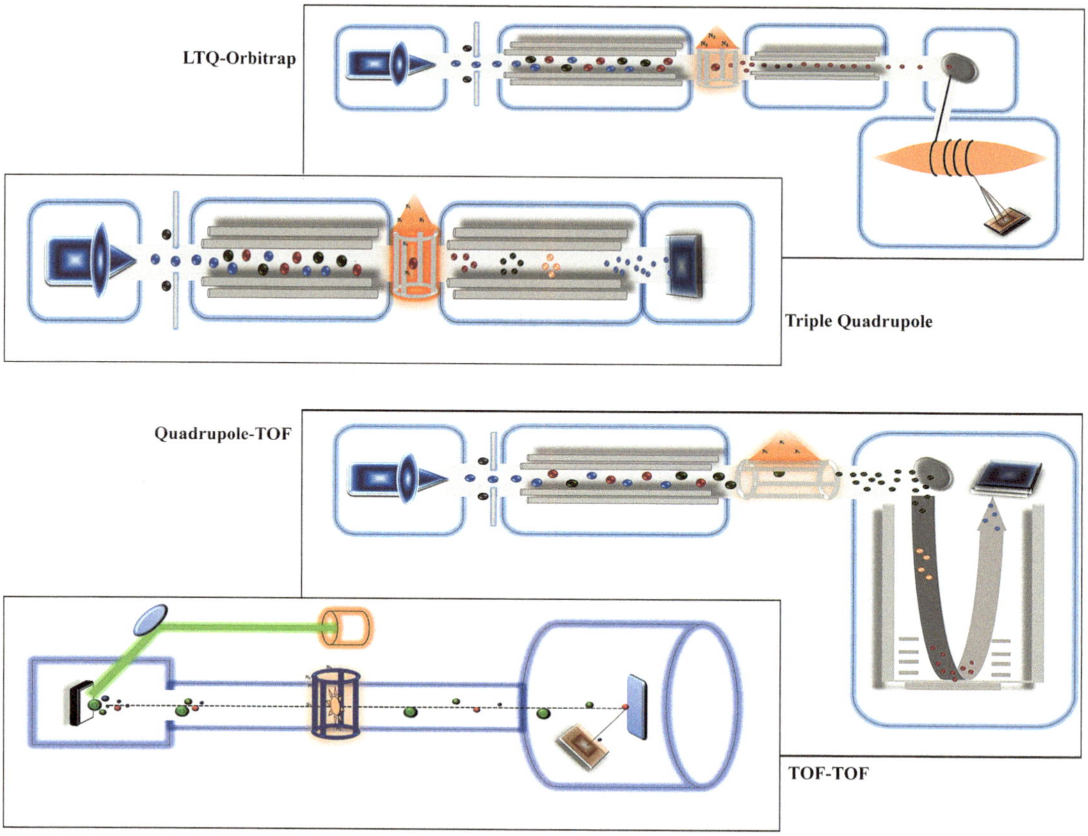

Fig. 4.1 Tandem mass spectrometry platforms. In tandem mass spectrometry, two mass analyzers in a series separated by collision cell. LTQ-Orbitrap, Triple quadrupole, quadrupole-TOF, and TOF/TOF are the widely used tandem mass spectrometry platforms for both shotgun and targeted proteomics

ation, and top-down sequencing which predominantly generates c and z ions. The third mass analyzer operates in "full-scan" mode where it allows all the fragmented ions and generates MS/MS spectrum for the selected peptide (Pitt 2009).

In targeted proteomic approach (SRM/MRM), the action of collision cell remains similar to that in shotgun proteomics with the difference being that the first and third mass analyzers operate in "radio frequency." The first mass analyzer selects the precursor ion, while the third mass analyzer measures the selected fragmented ion. This approach is best suited for quantification due to its high specificity, sensitivity, and dynamic range (MacLean et al. 2010). Though tandem mass spectrometry is an advanced and effective mass spectrometry technique, the complexity of the proteome has long posed a big challenge, thereby limiting the coverage. However, the emergence of better pre-fractionation methods over the last decade have helped in improving the coverage, though there still remains a lot of scope for development. The various pre-fractionation methods commonly employed have been described in the sections to follow.

4.1.3 Detector

The role of the detector in a mass spectrometer is to record the current produced or the charge induced when the ion beam from the mass analyzer hits the surface of detector. The selection of a detector depends on detection sensitivity, speed, stability, cost, collection efficiency, dynamic range, uniform response, and compactness. Different types of detectors currently in practice include the electron multiplier tube which measures the emission of secondary electrons generated by the ions, Faraday cup which generates current in vacuum when the charged particle strikes the suppressor electrode, and microchannel plate detector which records a weak AC image generated by oscillation of ions as they pass between the electrode plates (Neetu et al. 2012).

4.1.4 Bioinformatics Tools

Large amount of data generated from various proteomics platforms posed a great challenge in managing, analyzing, and interpreting subsequently, leading to the development of sophisticated bioinformatics tools for data management and analysis. The development of web-based open-source softwares owing to the generation of large amount of data has resulted in the emergence of a new era in proteomics-based research. Softwares such as Mascot, SEQUEST, and X!Tandem played a key role in the early years of this era paving way for the development of label-free and label-based quantitative proteome analysis (Kumar and Mann 2009; Deutsch et al. 2008).

4.1.5 Pre-fractionation Methods

Despite rapid advancements in the field of mass spectrometry, the complexity of clinical samples due to biological and technical limitations proves to be a major challenge to overcome, thereby compromising with the total proteome coverage. The biological limitation is a result of the masking of the less abundant proteins by highly abundant proteins in the samples, while the technical limitations are mainly due to limited reproducibility, sensitivity and resolution of the instrument, and chromatography platforms available.

Ultracentrifugation, the oldest and classical way for fractionating different organelles in tissue samples, cannot be applied to body fluids, thereby necessitating the need to explore other methods to decrease sample complexity. SDS-PAGE, which separates molecules based on their molecular weight, and 2-DE, which separates the molecules based on their isoelectric pH as well as molecular weight, have for long been used to decrease the complexity of samples. In addition, capillary electrophoresis which separates molecules based on charge and OFFGEL fractionation which separates proteins based on isoelectric pH in solution find widespread use in laboratories around the world (Li et al. 2005; Barnea et al. 2005). Many chromatographic methods have also emerged to reduce the complexity of the proteome both at protein or peptide levels. The separation principle of protein/peptide depends on the chromatographic technique being employed, with each technique exploiting a different property of the molecules to be separated. For example, affinity chromatography (iMAC, TiO_2, lectin) works based on affinity between the ligand and protein/peptide, ion exchange chromatography separates on the basis of charge on protein/peptide, size exclusion chromatography (sepharose, sephadex) separates on the size and shape of the protein, and reversed phase (RP)/hydrophobic chromatography exploits the hydrophobicity of protein/peptide. Despite the availability of a number of chromatographic methods, strong cationic exchange (SCX) chromatography is the method of choice in most laboratories. A combination of RP with SCX called as MudPIT (multidimensional protein identification technology) is another frequently used method yielding the best results among all other fractionation techniques available (Barnea et al. 2005; Pernemalm et al. 2009).

4.1.6 Mass Spectrometry-Based Proteomic Workflow

Protein extraction plays a crucial role in any proteomics experiment and the choice of protein extraction protocol employed is of great importance. Different protocols are usually applied depending on the sample under study. For example, in case of body fluids like blood, serum, urine, saliva and CSF the sample can directly be subjected to tryptic digestion (add 1 % SDS, reducing agents, and ammonium bicarbonate (ABC) to adjust pH to 8.0), or protein can be precipitated using TCA-acetone or acetone followed by dissolution of the protein pellet in ABC buffer followed by in-solution digestion. In case of protein extraction from tissues, microbes, or cell lines, the first and foremost step is efficient cell lysis using one of the techniques like sonication, enzymatic digestion, and liquid nitrogen grinding depending on the tissue or cell type. The cell lysate so obtained can be digested directly using trypsin or the protein may be precipitated with TCA-acetone or acetone and redissolved in dissolution buffer (ABC buffer) and subjected to tryptic digestion. High-resolution mass spectrometers are very sensitive to interfering agents like DNA, lipids, salts, etc., and may clog the column resulting in skewed results. So, clean-up (Zip-tipping using C18 columns) of samples is mandatory prior to injecting the samples for mass spectrometry analysis (Reddy et al. 2015a,b) (Fig. 4.1).

The workflow for shotgun proteomics using MS involves pre-fractionation followed by LC-MS/MS analysis, while the targeted proteomics workflow involves direct injection of the sample into LC-MS/MS. In case of SDS-PAGE & 2-DE pre-fractionation methods, the gel is cut into slices and in-gel digestion (using trypsin) is performed for subsequent LC-MS/MS analysis. In case of gel-free pre-fractionation, the proteins are digested in solution using trypsin (in-solution digestion) followed by fractionation using SCX, OFFGEL fractionation, RP chromatography, or iMAC. The fractionated peptides are subjected to liquid chromatography (having analytical and enrichment columns) connected online to tandem MS for identification and quantification. In case of targeted proteomics, the peptide mixture is directly subjected to online LC-separation having analytical and enrichment columns connected to tandem MS (Bantscheff et al. 2012).

4.1.7 Quantitative Proteomics Platforms

Conventional gel-based proteomic approaches (2-DE and DIGE) had been used in the past for quantitative proteomics but failed due to poor reproducibility, low sensitivity, and being limited to soluble proteins. Since the last decade, mass spectrometry-based quantitation has attained a lot of interests due to advancement in technology (Megger et al. 2013). Label-free quantification using mass spectrometry can be done either by spectral counting or by measuring the peptide peak intensity. In spectral counting, the number of peptides and its corresponding fragment ion (MS/MS) spectra are considered for relative quantification and identification. The number of peptides generated during MS and selection of the peptide for fragmentation is directly proportional to the protein abundance (low-abundant proteins often generate less peptides and fragmentation spectra). In peptide peak intensity, the relative quantification of the protein/peptide can be performed by measuring the peptide peak intensity at MS level and integrated with corresponding chromatography peak intensity. Label-free quantification is fast, simple, and cost-effective and has high dynamic range but does not allow multiplexing and is less reproducible (Wasinger et al. 2013). Label-based quantification methods such as iTRAQ, TMT, iCAT, SILAC, etc., overcome the limitations of label-free methods. It is to be noted that iTRAQ and TMT labels can be used for in vitro labeling of clinical samples, whereas SILAC labeling can be exclusively applied to cell cultures since it uses labeled amino acids for labeling (Fig. 4.2). 8-plex iTRAQ quantitative proteome analysis performed by labeling the proteins from different stages of developing human embryonic stem cells yielded 156 differentially expressed proteins involved in cell proliferation, apoptosis, and protein synthesis (Jadaliha et al. 2012). Similarly, biomarker discovery in periodontal disease was performed using TMT labeling leading to the conclusion

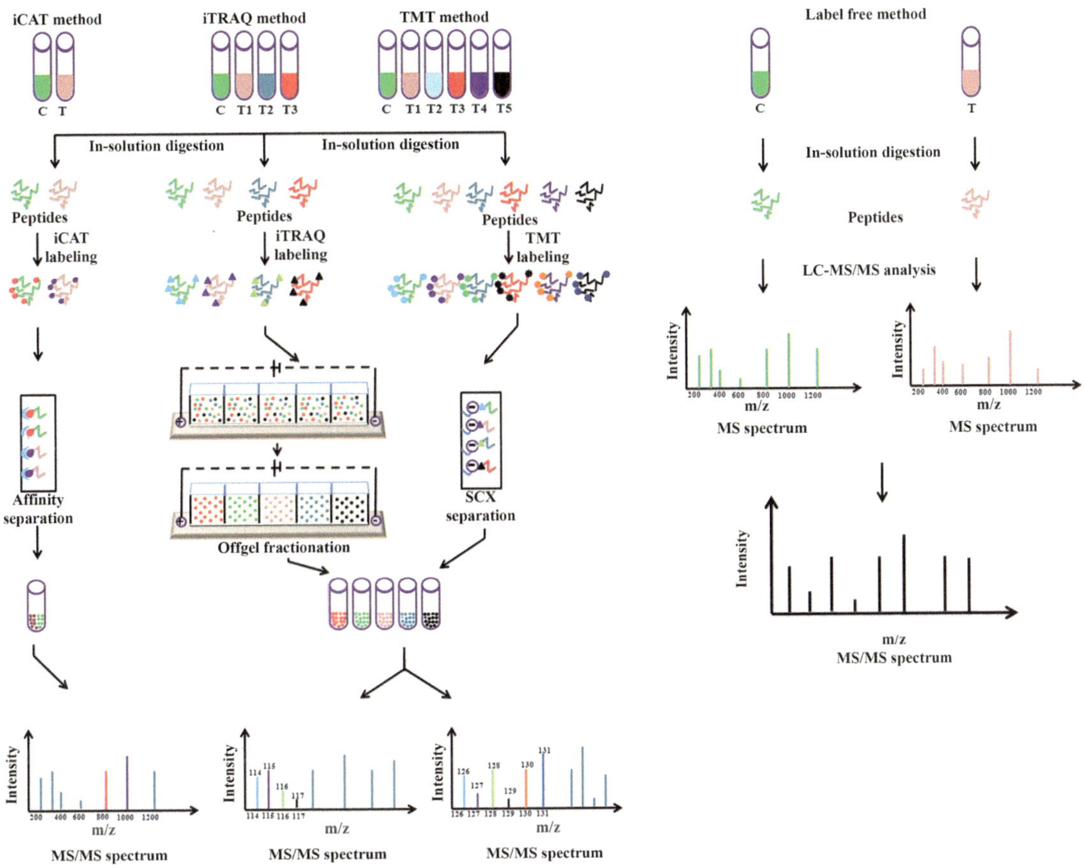

Fig. 4.2 Quantitative proteomics workflow. The standard procedure for LC-MS/MS-based quantitative proteomic analysis using iCAT, iTRAQ, TMT, and label-free quantification. The complex biological samples are fractioned and analyzed with LC-MS/MS analysis in case of iCAT, iTRAQ, and TMT. In case of label-free quantification, direct sample injection is preferable

that matrix metalloproteinase-9 and neutrophil gelatinase-associated lipocalin can serve as potential markers for disease progression (Tsuchida et al. 2013).

4.2 Biomarker

A Biomarker may be a biomolecule from–complex biological fluids or tissues which can differentiate two closely related states of the biological system. A biomarker can be a protein, nucleic acid, lipid, a receptor, antibody, peptide, and glycan or can also be a change in expression of metabolites, proteins, or genes produced as a result of a diseased or unhealthy state. Over the last two decades, a number of articles have been published about the role of thousands of biomarkers in different diseases. However, not many of these published biomarkers have been put to use in the medical field since most studies either select the wrong candidates in the discovery phase or lack precise validation/characterization or may only be detected using very sensitive techniques which are otherwise not easily available or affordable in clinical settings (Henry and Hayes 2012).

There are different types of biomarkers such as diagnostic biomarkers, prognostic biomarkers, predictive biomarkers, and pharmacodynamic biomarkers. Molecular markers useful in identification of a specific disease are called diagnostic biomarkers. For example, elevated levels of prostate-specific antigen (PSA) in serum are used as a diagnostic marker in detecting/diagnosing prostate cancer (Coley et al. 1997). Prognostic biomarkers are used for monitoring disease progression and assessing the final outcome of the disease (good/poor prognosis). Serum VEGF level is used as a prognostic marker in renal cell carcinoma (Oldenhuis et al. 2008). Predictive biomarkers give an idea about the patients who are benefited by a specific drug treatment. In case of breast cancer, the patients with overexpression of ERBB2 gene respond to the treatment with trastuzumab (Herceptin), whereas the patients with high expression levels of estrogen receptors by the tumor cells respond well to the treatment with tamoxifen (Sawyers CL 2008). Pharmacodynamic biomarkers are helpful in determining the drug dosage to treat the patients more effectively. BCR-ABL kinases act as pharmacodynamic biomarkers, whose activity determines dosage of the drug (Dasatinib) that needs to be administered to chronic myeloid leukemia (CML) patients (Shah et al. 2006). Discovery of multiple biomarkers helps in proper diagnosis of the disease and helps the clinicians to select the kind of therapy which benefits the patients the most.

4.3 Application of Mass Spectrometry in Cancer Biomarker Discovery

The number of cancer incidences has been increasing every year around the world because of the sedentary lifestyle, smoking, unhealthy food habits, and genetic factors. Early diagnosis and proper treatment can reduce the mortality rate significantly. New and efficient cancer biomarkers are hence of utmost importance and the quest for the cancer biomarkers started with the discovery of tumor-produced antibody light chains in the urine of multiple myeloma patient by Henry Bence-Jones in 1847 (Bence-Jones 1847). Since then, only 23 proteins have been approved by FDA as cancer biomarkers and are currently being used in clinics (Füzéry et al. 2013). Most of the FDA-approved proteins however are either not so sensitive or specific to a particular cancer (Polanski and Anderson 2007). So we need a panel of protein markers, viz., biosignatures which could accurately discriminate cancer from the healthy state.

4.3.1 Prostate Cancer

Among men, prostate cancer is the second most common cancer and sixth leading cause of cancer deaths (Jain et al. 2014). Quite a few biomarkers for prostate cancer are currently in use. Papsidero et al. reported elevated levels of serum prostate-specific antigen (PSA) in prostatic cancer patients, thereby concluding it to be a potential biomarker for prostate cancer (Papsidero et al. 1980). However, serum levels of PSA were found to increase in other clinical conditions like prostate irritation, prostatic infection, and benign prostatic hyperplasia (BPH) (Arneth 2009). In many cases, serum PSA levels in prostate cancer patients were observed to be in the normal range. Such cases often create enough ambiguity about the patient's status in a clinician's mind. Hence, the need for a more specific biomarker/s is imperative for diagnosis and prognosis of prostate cancer. An interesting study by Michiko et al. came up with a novel biomarker for prostate cancer, thus giving a better picture of a patient's state with elevated levels of PSA and improving the treatment possibilities to a greater extent. 2DICAL (2-Dimensional Image Converted Analysis of Liquid chromatography and mass spectrometry), a quantitative label-free shotgun proteomics platform, was used in a study which revealed elevated levels of carbonic anhydrase I as a novel plasma biomarker for prostate cancer. The findings were further validated by immunological assays (Takakura et al. 2012). Grover et al. observed the presence of 22 kDa protein in prostate cancer patients when compared with normal subjects and benign prostatic hyperplasia patients. The protein was also found to be excreted in the urine of prostate cancer patients (Grover and Resnick 1995). Another study reported by Hwang et al. described about a novel biomarker for identifying prostate cancer using prostate tissue samples. The study used DTP (direct tissue proteomics) coupled with LC-MS/MS wherein the prostate tissue samples were fixed using formalin and embedded in paraffin, followed by identification of proteins by mass spectrometry. The findings of the study introduced another novel biomarker,Wnt-3 that is upregulated in prostate cancer cells. The combined use of PSA and Wnt3 as biomarkers for identification of prostate cancer will be more successful for prognosis and diagnosis cases of prostate cancer (Hwang et al. 2007).

4.3.2 Ovarian Cancer

Among all the cancers in women, ovarian cancer is the seventh most common across the world (Zhang et al. 2000). Most of the ovarian cancer patients are diagnosed at their late clinical stage, and due to this, 35 % of patients have a survival period of 5 years, whereas 90 % of the ovarian cancer patients diagnosed at the earlier stages have a survival period of 5 years (Menon and Jacobs 2000; Cannistra 2004). Serum CA125 is commonly used as a biomarker for ovarian cancers. Elevated levels of serum CA125 have been observed in 80 % of the ovarian cancer in advanced stages and 50–60 % elevation in the early stages. Zhang et al. analyzed the serum proteome from healthy and early stage ovarian cancer patients using SDS-PAGE followed by mass spectrometry. Three acute phase proteins (apolipoprotein A1, truncated form of transthyretin and inter-α-trypsin inhibitor heavy chain H4) identified by Zhang et al. were found to be useful in the detection of early stage ovarian cancer (Zhang et al. 2004). In another study, Mary et al. discussed ways to identify biomarkers to enable detection of ovarian cancer at an early stage of the disease onset to prevent high rate of mortality. The method employed to study this was to enrich serum proteins by carrier protein-bound affinity column, followed by MALDI orthogonal TOF mass spectrometry analysis. A series of biomarker panels were identified successfully, namely, Keratin 2a, complement component 3 precursor, complement component 4A preprotein, casein kinase II, alpha 1 subunit isoform a, inter-alpha (globulin) inhibitor H4, fibrinogen, and few more (Lopez et al. 2007). Using ELISA- and mass spectrometry-based approaches (Both SELDI-MS and LC-MS/MS), it was observed that the serum levels of alpha chain of haptoglobin increased in ovarian cancer patients. Alpha chain of haptoglobin alone had 64 % sensitivity and 90 % specificity for detecting the ovarian cancers, whereas haptoglobin alpha chain along with CA125 had 91 % sensitivity and 95 % specificity (Ye et al. 2003).

4.3.3 Lung Cancer

Lung cancer is the leading cause of cancer deaths around the world. In most cases, lung cancers are diagnosed in the advanced stages, due to which only 15 % of these patients have a 5-year survival time (Reddy et al. 2011). Increased serum levels of haptoglobin were observed in the lung cancer patients using SDS-PAGE coupled with LC-MS/MS analysis (Kang et al. 2011). Howard et al. identified significant increase in the serum levels of serum amyloid A (SAA) protein in lung cancer patients when compared to healthy subjects using MALDI-TOF mass spectrometry. The increased levels of SAA were validated using ELISA (Howard et al. 2003). Heo et al. used multi-lectin affinity columns for the enrichment of the glycoproteins from serum of both lung cancer patients and healthy controls and separated on SDS-PAGE. The LC-MS/MS analysis of each band from the gel showed an increase in serum levels of plasma kallikrein (KLKB1) and inter-α-trypsin inhibitor heavy chain 3 (ITI-H3) in lung cancer patients (Heo et al. 2007). In another study performed by Zeng et al. employing label-free quantification using LC-MS/MS analysis, 22 serum glycoproteins were reported to be altered in NSCLC (Zeng et al. 2010). Proteins associated with inflammatory response, cell-cell signaling, and various interaction networks were dysregulated in the non-small-cell lung cancers (NSCLC), which could be useful in the diagnosis of early stage lung cancers (Zeng et al. 2011). Campa et al. showed increased expression of macrophage migration inhibitory factor and cyclophilin A in lung cancer tissues as a potential diagnostic marker (Campa et al. 2003). Chris et al. selected four histological different lung cancer cell lines, viz., adenocarcinoma, squamous cell carcinoma, and small-cell and non-small-cell lung cancer cell lines, and analyzed using bottom-up proteomics approach. The potential markers obtained from cell lines were validated in both healthy and lung cancer patient samples, and the study revealed five novel biomarkers for lung cancer, namely, ADAM-17, osteoprotegerin, pentraxin 3, follistatin, and tumor necrosis factor receptor super family member 1A (Planque et al. 2009).

4.3.4 Breast Cancer

Breast cancer is the second leading cause of cancer deaths in women all over the world (Abdulkareem 2013). Mammography, ultrasound, and histopathology have since long been used as diagnostic tools for the breast cancer (Chung et al. 2014). However, there is a need for discovery of biomarkers for quick and early diagnosis of breast cancer. Overexpression of HER-2/neuoncoprotein is used as a biomarker for the detection of breast cancers, but there still exists a need for a more specific biomarker (Têtu and Brisson 1994). Geiger et al. identified high levels of IDH2 and CRABP2 and low levels of SEC14L2 in the ER-negative breast cancer tumors using mass spectrometry-based approach. The expression of IDH2, CRABP2, and SEC14L2 were found to be associated with the overall survival rate of the patients. IDH2 and CRABP2 offered poor prognosis, while SEC14L2 offered good prognosis (Geiger et al. 2012). Profumo et al., using MALDI-MS-based approach, studied the serum proteome of the gross cystic disease (GCDB), a benign disease that later develops into breast cancer, and identified complement C3f levels as a marker protein for predicting the risk of breast cancer in GCDB patients (Profumo et al. 2013). Using LC-MS-based approach, Villanueva et al. identified 14 potential serum biomarker peptides for breast cancers, of which 11 peptide markers were unique to breast cancer only and could discriminate breast cancers from healthy subjects and other cancers (Villanueva et al. 2006). Van den Broek et al. quantified the concentrations of six serum peptides such as bradykinin, Hyp3-bradykinin, des-Arg9-bradykinin, fragments of fibrinogen α-chain (Fib-α[605–629]), inter-α-trypsin inhibitor heavy chain 4 (ITI-H4), and complement component 4a (C4a) in breast cancer patients using targeted MRM-based mass spectrometric approach (van den Broek et al. 2010). Vathany et al. performed bottom-up proteomics approach using 2D LC-MS/MS on MCF-10A, BT474, and MDA-MB-468 cell lines. Of the many differentially expressed proteins, elafin, a protease inhibitor, and three kallikreins, viz., KLK5, KLK6, and KLK10, were reported as potential biomarkers for breast cancer following their validation using immunoassays in serum and tissue samples (Kulasingam & Diamandis2007).

In addition to the above mentioned cancers, work is in progress for the discovery of biomarkers for other cancers. Protein levels of serum paraoxonase, transthyretin, annexin VI, serum amyloid A, ceruloplasmin, etc., were found to be altered in hepatocellular carcinoma (Yang et al. 2007). A panel of three proteins, apolipoprotein A-II, transthyretin, and apolipoprotein A-I, were identified using MALDI-TOF MS method in the serum samples of pancreatic cancer patients (Ehmann et al. 2007). Using LC-MS-based approach, Hsu et al. identified five urinary nucleosides to be elevated in colorectal cancer that could discriminate colorectal cancer patients from the healthy subjects (Hsu et al. 2009). Proteomic analysis of childhood acute lymphoblastic leukemia cell proteome with the normal peripheral lymphocytic cells revealed upregulation of glutathione S-transferase P and prohibitin in acute lymphoblastic leukemia patients substantiating their potential as diagnostic markers (Wang et al. 2012). Another study by Findeisen et al. revealed that increased serum levels of serum amyloid A along with the C-reactive proteins could be used as prognostic marker for early stage melanoma patients (Findeisen et al. 2009). Using MALDI-

TOF MS approach, Yang et al. identified RNA-binding protein 6, tubulin beta chain, and zinc finger protein 3 in renal cell carcinoma and hold potential for use as diagnostic markers for renal cell carcinoma (Yang et al. 2014).

4.4 Applications of MS in Biomarker Discovery in Other Diseases

4.4.1 Infectious Diseases Biomarker

Infectious diseases are the world's second leading health concern and a recent report has highlighted that more than 1,400 microbial human pathogens have been detected so far with the number increasing progressively (Hay et al. 2013). Initially, next-generation sequencing helped to diagnose the infectious diseases but failed to continue due to technical and practical problems with culturing of the microbes (Lecuit and Eloit 2014). Over the last decade, mass spectrometry-based proteomics has been extensively used to explore biomarkers in infectious diseases for early diagnosis and prognosis by analyzing the host proteome. Body fluids (serum, plasma, saliva, CSF) are the better biological samples for examining infectious diseases. In some cases, tissue-specific proteomics provide fruitful information to understand disease pathogenesis. Recently, Wang et al. (2014) reported rapid laboratory diagnosis method for respiratory infection pathogen by using MALDI-TOF-MS as a fast, reliable, and inexpensive method (Wang and Fu 2014). Tuberculosis (TB) is a dreadful infectious disease with high mortalityand limited diagnosis. iTRAQ-based 2D LC-MS/MS analysis of serum from pulmonary TB, lung disease, and healthy samples resulted in a panel of three proteins, namely, S100A9, SOD3, and MMP9, having an ability to clearly differentiate TB from other lung diseases with good specificity and sensitivity (Zhou 2015). A study employing SELDI-TOF MS in combination with weak cation exchange magnetic beads for TB serum biomarker detection showed that a fragment of fibrinogen and alpha polypeptide isoform alpha-E preproprotein could be a potential marker for TB diagnosis (Liu et al. 2013a). Kruh-Garcia et al. used MRM-based mass spectrometry analysis to detect Mtb (*Mycobacterium tuberculosis*) proteins in the exosomes from infected human serum. The study reported 20 bacterial proteins in the exosomes with need for further validation in order to consider them as biomarkers for diagnosis in the future (Kruh-Garcia et al. 2014). Malaria is the leading infectious disease in the tropical and subtropical region with a million deaths reported worldwide every year. Comparative LC-MS/MS-based analysis of serum from malaria patients and healthy individuals revealed differential expression of apolipoprotein E (ApoE), serum amyloid A (SAA), gelsolin, complement factor H, fibrinogen, and lipid-binding protein. It was also observed that these proteins had ability to bind to hemozoin, a protein with a role in malarial disease progression (Kassa et al. 2011). The saliva proteome analysis from malaria patients identified *P. falciparum* proteins such as PFL0480w, PF08_0054, and PFI0875w in addition to upregulation of erythrocyte and inflammatory proteins (Huang et al. 2012). An independent study on serum from malaria using iTRAQ-based quantitative proteome analysis reported increased levels of cell adhesion molecule-4 and C-reactive protein and decreased levels of haptoglobin (Mu et al. 2014).

4.4.2 Diabetes Mellitus Biomarker

Diabetes mellitus (DM) is a chronic metabolic disorder affecting nearly 5 % of the world population with its incidence increasing every year. The characteristic feature of this disease is the persistent hyperglycemic condition in the patients, leading to major complications such as retinopathy, nephropathy, and neuropathy. The measurement of glycated hemoglobin is a classical way to detect and monitor DM. Proteome and metabolome-based biomarkers have gained attention in many diseases owing to the need to detect, depict, and prevent the progression of the disease. The salivary proteome analysis of type-II DM and healthy individuals using 2D-LC-

MS/MS analysis revealed majority of the altered proteins to be involved in metabolic process and immune response (Rao et al. 2009). Similarly, serum proteome analysis using iTRAQ-based quantitative proteomic analysis (LC-MS/MS) by Kaur et al. revealed alteration of quite a few proteins involved in metabolism and other cellular processes. MRM-based validation of apolipoprotein-A1, afamin, transthyretin, vitamin D-binding protein, and fibronectin showed their involvement in DM (Kaur et al. 2012). Mass spectrometry-based metabolomics is one of the finest ways to study whole metabolites in biological samples. Metabolome analysis of blood samples from diabetes and non-diabetes using tandem mass spectrometry and nuclear magnetic resonance (NMR) spectroscopy revealed a number of altered metabolites (known and novel metabolites) with the ability to differentiate DM from healthy individuals. Differentially expressed metabolites include sugar, ketone body, and branched-chain amino acids (Suhre et al. 2010). An independent study showed acetone, a ketone body synthesized in DM as a potential noninvasive biomarker. The group measured the concentration of acetone in breath from both DM and healthy individuals using low-temperature cofired ceramics (LTCC) technology coupled with mass spectrometry operated in multiple ion detection mode by selecting the m/z as 43 and 58 (for acetone) (Rydosz 2014).

4.4.3 Cardiovascular Diseases Biomarker

Cardiovascular disease (CVD) is the leading cause of death worldwide. The subclinical complications and risk factors of CVD are very severe and life-threatening. CVDs are quite common in developed countries and the incidents are increasing every year. The classical diagnostic methods such as electrocardiogram, stress testing, echocardiography, chest X-ray, blood test, and natriuretic peptides are effective but are limited by their inability to diagnose the disease in the early stages. Omics-based (proteomic and metabolomic) diagnostic markers are hence indispensable for early diagnosis and for monitoring disease progression. LC-MS/MS-based quantitative proteome and metabolome of serum/urine or the heart tissue could provide useful information to monitor CVD. Magiera et al. quantified the concentration of α-ketoglutaric acid (α-KG), L-carnitine, and acetyl-L-carnitine in urine samples collected from CVD using calibration standards prepared with purified small molecules in LC-MS/MS analysis and reported their potential as biomarkers for diagnosis of CVD (Magiera et al. 2013). COmbined FRActional DIagonal Chromatography (COFRADIC) in combination with LC-MS/MS proteomic platform was used for studying the plasma proteome of acute decompensated heart failure (ADHF), and quiescin Q6 was reported as a potential marker for ADHF from the study (Mebazaa et al. 2012). Another study on coronary heart disease (CHD) using iTRAQ-based LC-MS/MS analysis revealed differential expression of 12 proteins (five upregulated and seven downregulated). The upregulated proteins were found to be involved in inflammation and downregulated proteins in lipid metabolism. Serum amyloid A (SAA) and apoC-I were further validated using ELISA and proved to beuseful in monitoring the disease (Yan et al. 2014). Similarly, the salivary proteome analysis of CVD was performed using weak cation exchange magnetic bead kit followed by identification with UltraFlex III MALDI-TOF/TOF. The analysis revealed differential expression of 11 peptides (five upregulation and six downregulation), and alpha-2-HS-glycoprotein was validated in saliva as a potential marker for disease diagnosis (Zheng et al. 2014).

4.5 Oncometabolomic Biomarker

Cancer is largely a genetic disorder with dysregulation of physiological metabolism being one of the reasons. Metabolites are small molecular weight key intermediates or end products of metabolism and regulate the phenotype of the system under genetic or environmental conditions. Monitoring metabolites in biofluids (serum, saliva, or urine) using biochemical reactions is a common practice in many pathophysiological conditions to diagnose or predict diseases (Gupta and Chawla 2013). For many years, radio pulse labeling was used to monitor the metabolites, but advancements in techniques such as NMR spectroscopy, GC-MS, and LC-MS have led to increased use of these high-throughput technologies in monitoring metabolites. The emergence of metabolomics as a new platform for high-throughput study of metabolites has greatly helped in increasing our understanding of physiology. NMR- and mass spectrometry-based (coupled with GC and LC) analysis is widely used for metabolome analysis with the help of advanced software and statistical tools (Zhang et al. 2013; He et al. 2014; Chan et al. 2009). The recent advances in technologies have enabled a better understanding of cancer metabolism and tumor proliferation with a lot of emphasis on discovering diagnostic biomarkers for cancers.

Biological fluids (serum, blood, plasma, saliva, urine, and cerebrospinal fluid), biological tissue, or cells can be used for metabolome analysis. The sample preparation is quite simple for biological fluids, whereas for tissue or cells, it is tedious due to contamination of biofluids or surrounding tissue. Metabolomic profiling of urine from women cancer patients (breast, ovarian, and cervical cancer) using mass spectrometry (GC-MS and LC-MS) revealed the metabolites involved in oxidative DNA damage and methylation process. The authors reported 5-hydroxymethyl-2-deoxyuridine and 8-hydroxy-2-deoxyguanosine, 1-methyladenosine, 3-methyluridine, and 4-androstene-3,17-dione as novel potential biomarkers in both breast and ovarian cancers (Woo et al. 2009). Qualitative metabolomic profiling of hepatocellular carcinoma (HCC) in Egyptian cohort was performed using UPLC-QTOF-MS, followed by targeted analysis of seven metabolites such as Phe-Phe, 3β,6β-dihydroxy-5 β-cholan-24-oic acid, GCA, GDCA, TCDCA, oleoylcarnitine, and linoelaidylcarnitine in 89 samples using UPLC-triple quadrupole linear ion trap. Among the seven potential biomarkers discovered, three biomarkers were already reported in HCC and four were unique to this study (Xiao et al. 2012). Another independent study involving metabolomic analysis of HCC (183 serum samples – 77 sera as discovery set and 106 sera as validation set) using capillary electrophoresis-mass spectrometry, PCA, and discrimination analysis revealed four out of six metabolites (tryptophan (Trp), pipecolic acid, glutamine (Gln), 3/4-methyl-2-oxovaleric acid, arginine (Arg), and 2-hydroxybutyric acid) to be able to discriminate small HCC (Zeng et al. 2014).

4.6 PTM Analysis Using Mass Spectrometry

Prior to gaining a clear insight into protein chemistry, it was believed that proteins formed after translation directly control physiological pathways. However, a better understanding of the dynamic nature of proteins brought to light the importance and role of various post-translational modifications in control of biological functions. Nearly 300 PTMs are possible on proteins in biological system with most of the biomarkers reported in various diseases being phospho- or glycoproteins (Olsen and Mann 2013). Though many PTMs exist, phosphorylation (one-third of eukaryotic proteins) and glycosylation (nearly 50 % of human proteins) are predominant PTMs in eukaryotes including human. Mass spectrometry has emerged as a central tool for evaluating site mapping and quantification of protein modifications (PTMs). Advancements in mass spectrometry (ETD or ECD fragmentation) have helped study the large-scale PTMs in complex samples (Fig. 4.3). In addition to this, sophisticated bioinformatic tools to analyze the data and establishment of PTM databases have greatly helped in our understanding of PTMs. The emergence of new fragmentation technologies and advanced tools has helped in PTM detection at both MS and MS/MS level. For example, at MS level, phosphorylation leads to increase in mass of amino acid by +80 Da (addition of HPO_3 on phosphorylated amino acid), whereas at MS/MS level, phosphorylation either leads to increase in mass on amino acid by +80 Da (addition of HPO_3

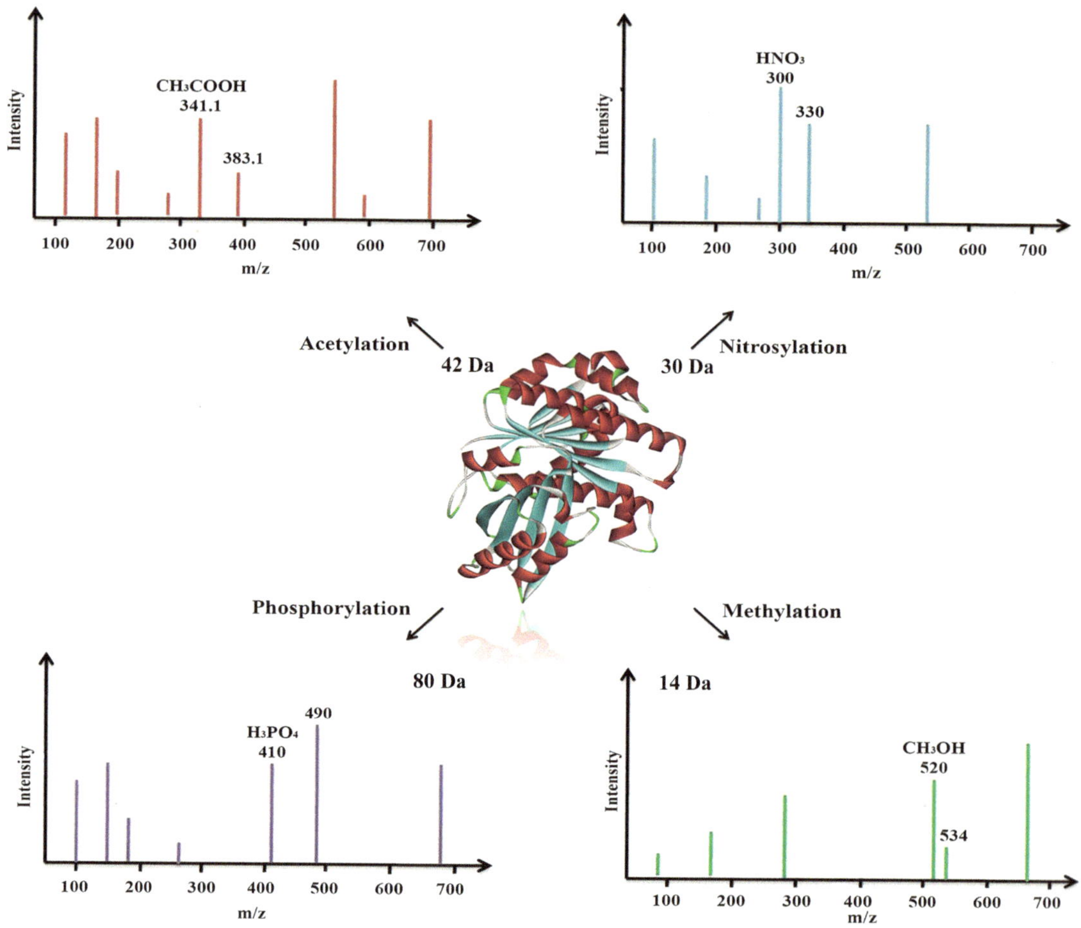

Fig. 4.3 Posttranslational modification (PTM) analysis. Mass spectrometry is the one of the best methods for PTMs such as phosphorylation, methylation, acetylation, nitrosylation, and so on. Here the spectra represent the PTMs on protein/peptide

on phosphorylated amino acid) or a decrease in mass by −18 Da on amino acid (removal of H_3PO_4 from phosphorylated amino acid leads to dehydrated amino acid) (Witze et al. 2007). In addition to this, enrichment technologies such as iMAC (immobilized metal affinity chromatography), titanium dioxide (TiO_2) and zirconium dioxide (ZrO_2) for phosphopeptides and lectin columns, and hydrazide chemistry for glycopeptides are increasingly being used to enrich the low-abundant PTM peptides (Zhang et al. 2014). The advantages of MS-based PTM analysis include differential expression analysis, high-throughput identification, modified site evaluation, quantification (absolute/relative), high sensitivity, and specificity, but the major limitation is the lack of sufficient and reliable databases for most PTMs. Choi et al. have used sequential enrichment columns with Fe_3O_4, followed by TiO_2 particles for phosphopeptide enrichment from digested α-casein, and the same method has been used for mono- and multi-phosphopeptide analysis from complex L6 muscle cell lysate using LC-MS/MS as a proof of concept (Choi et al. 2010). The protein extracted from leukocytes was used for phospho-analysis after fractionation with SCX, followed by LC-MS/MS analysis using phosphochip. The instrument was calibrated with three test phosphopeptide samples (BSA, α-casein, and β-casein) prior to the leukocyte sample analysis. More than 250 phosphopeptides were detected with good reproducibility among triplicates (Raijmakers et al. 2010). In another study, sera from lung cancer (depleted) were passed through hydrazide resin, and in situ trypsin digestion was performed to remove the non-glycosylated peptides. A total of 38 glycopeptides (representing 22 proteins) were detected using tandem MS/MS analysis after eluting with peptide-N-glycosidase F. Three glycoproteins, α-1-antichymotrypsin, insulin-like growth factor-binding protein 3, and lipocalin-type prostaglandin D synthase, were validated using ELISA and considered as potential biomarker for lung cancer (Zeng et al. 2010). SWATH, the data independent analysis of tissue from prostate cancer (PC), revealed differential expression of 220 glycoproteins involved in aggression and metastasis of PC. Further, validation of PC using tissue microarray highlighted two glycoproteins, namely, N-acylethanolamine acid amidase and protein tyrosine kinase 7 as potential markers to monitor the aggression of PC (Liu et al. 2014).

4.7 Targeted Proteomics Assays Using Mass Spectrometry

"Shotgun proteomics" or "discovery proteomics" using high-resolution mass spectrometry coupled with liquid chromatography has revealed many potential biomarkers in various diseases. Validation of these biomarkers is a tedious process due to expensive antibody-based immune assays, limitation of antibody availability, and time-consuming protocols (Schiess et al. 2009). Absolute quantification of protein is of prime importance in clinical biochemistry for diagnosis. MS-based targeted proteomics, i.e., multiple reaction monitoring (MRM) or selected reaction monitoring (SRM), are the alternatives and the data is reproducible, and of high sensitivity, specificity, and dynamic range (McIntosh and Fitzgibbon 2009). The principle of SRM/MRM involves the selection of both precursor (Q1) and fragmented ion (Q2), and measuring the peak intensity of Q1/Q3 transition gives the quantification of the peptide/protein. Triple quadrupole (QqQ) instruments yield best results for SRM/MRM, where both Q1 and Q3 operate in radio frequency mode to select the specific precursor and fragmented ions and collision cell (q2) in between the Q1 and Q3 (Colangelo et al. 2013) (Fig. 4.4). In MRM/SRM analysis, the selected peptide should be unique to the protein, should not be modified, should have high intensity, should give good fragmentation, and should have synthesized heavy peptide, and instrumental parameters should be optimized. Skyline is one of the universal tools used to persuade all the above criteria for SRM/MRM analysis across the globe (MacLean et al. 2010). Percy et al. tested 27 potential marker proteins in breast cancer using MRM-based assay. For each protein, one to four transitions have been studied and 82 heavy peptides were synthesized (internal standards) to quantify the protein in the plasma samples. In undepleted serum sample, the lowest detection range was 127 ng/mL for insulin-like growth factor (Percy et al. 2013). Similarly, anterior gradient 2 (AGR2) protein involved in epithelial cancer was measured in serum and urine after fractionation with PRISM, followed by LC-SRM analysis. The limit of quantification was ~130 pg/mL in serum and ~10 pg/100 μg in urine, and data reproducibility was high and also showed good correlation with ELISA results (Shi et al. 2014).

4.8 Mass Spectrometry Imaging for Cancer Biomarker Discovery

An image undoubtedly conveys deeper information to understand biology at system level. The conventional imaging techniques such as angiography, CT scanning, EEG, and MRI have been used more frequently in the bio-medical sector. However,they have one or few other limitations which lead to narrow spectrum usage, tremendous data loss, and inefficiency in acquiring certain sets of data. Imaging mass spectrometry

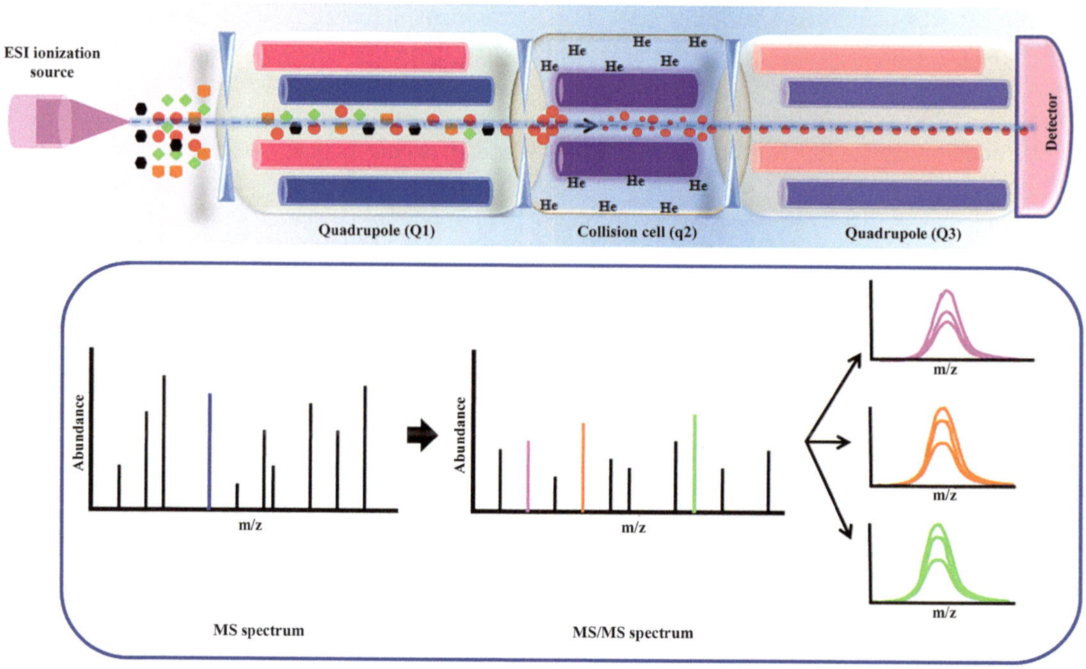

Fig. 4.4 Targeted proteomics workflow. Triple quadrupole (QqQ) is the preferred mass spectrometry for SRM/MRM analysis. Both Q1 and Q3 operate in radio frequency mode to quantify the selected precursor (peptide) and fragmented ion and the quantification of the protein/peptide obtained from Q1/Q3 ratio

(IMS) is a technique to visualize the spatial distribution of peptides, proteins, lipids, metabolites, etc., through a tissue to understand anatomy. Matrix-assisted laser desorption/ionization (MALDI) MS is the best platform for imaging, and the principle involves sectioning of the tissues into small slices (5–10 μm) and coating with matrix (Cornett et al. 2007). A grid can be made according to the tissue slice and data can be acquired with predefined laser shots in regular patterns on each grid coordinate (Schwamborn 2012). The final data contains a range of spectra in which each spectrum represents an independent molecular profile of the different area within the sample and integration of the entire data provides the anatomical view (Fig. 4.5). The advantages of IMS include lack of labels or tags, providing direct correlation with anatomy and high-throughput data. The lack of direct validation method is the sole limitation in taking IMS to the clinical level. One of the major applications of IMS is in cancer biomarker discovery. IMS has been used to study the molecular structure of brain tumor in mice model. Tumor tissues generate different imaging patterns depending on locations and stage of diseases indicating the heterogeneity of the tumor (Ait-Belkacem et al. 2014). IMS imaging analysis of prostate cancer tissue has revealed higher content of lipid components such as 14 phosphatidylinositols (PIs), 3 phosphatidyl ethanolamines (PEs), and 3 phosphatidic acids (PAs). The high expression of PI profiles could serve as good molecular markers for diagnosis of prostate cancer (Goto et al. 2014).

4.9 Conclusions

In the present chapter, we focused on the basics of mass spectrometry and its advancements in last decade. We also discussed the application of mass spectrometry in biomarker discovery in major diseases in the world such as cancer,

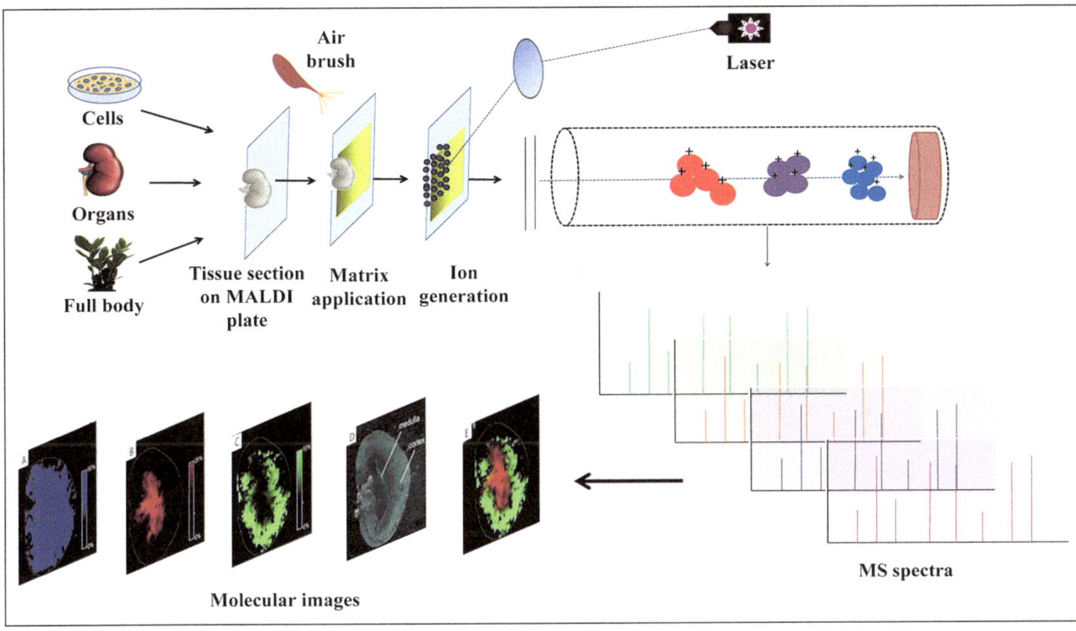

Fig. 4.5 MALDI imaging workflow. MALDI-TOF is the ideal instrument for tissue imaging. The tissues were sectioned into thin slices and sprayed with matrix. The data acquisition was performed at different parts of the tissue and integration of whole provides the molecular image of the tissue

diabetes, cardiovascular, and infectious diseases. In addition to biomarker discovery, mass spectrometry has wide range of applications in proteomics and is one of the best methods for PTM analysis at present. The mass spectrometry field is growing tremendously and is also playing crucial role in metabolomics and tissue imaging. The high-resolution mass spectrometry has also contributed significantly to the first human proteome draft and continuously evolving.

References

Abdulkareem IH (2013) Aetio-pathogenesis of breast cancer. Niger Med J 54:371–375. doi:10.4103/0300-1652.126284

Aebersold R, Mann M (2003) Mass spectrometry-based proteomics. Nature 422:198–207. doi:10.1038/nature01511

Ait-Belkacem R, Berenguer C, Villard C et al (2014) MALDI imaging and in-source decay for top-down characterization of glioblastoma. Proteomics 14:1290–1301. doi:10.1002/pmic.201300329

Arneth BM (2009) Clinical significance of measuring prostate specific antigen. Lab Med 40:487–491

Bantscheff M, Lemeer S, Savitski MM et al (2012) Quantitative mass spectrometry in proteomics: critical review update from 2007 to the present. Anal Bioanal Chem 404:939–965. doi:10.1007/s00216-012-6203-4

Barnea E, Sorkin R, Ziv T et al (2005) Evaluation of prefractionation methods as a preparatory step for multidimensional based chromatography of serum proteins. Proteomics 5:3367–3375. doi:10.1002/pmic.200401221

Bence-Sones H (1847) Papers on chemical pathology: lecture III. Lancet ii:26F–72F

Boylan KL, Andersen JD, Andersen LB et al (2010) Quantitative proteomic analysis by iTRAQ(R) for the identification of candidate biomarkers in ovarian cancer serum. Proteome Sci 8:31. doi:10.1186/1477-5956-8-31

Busto ME, Montes-Bayon M, Anon E et al (2008) Simultaneous determination of glycated haemoglobin, a long term biomarker of diabetes mellitus, and total haemoglobin by isotope dilution and HPLC-ICP-MS. J Anal Atom Spectrom 23(5):758. doi:10.1039/B718008J

Campa MJ, Wang MZ, Howard B et al (2003) Protein expression profiling identifies macrophage migration inhibitory factor and cyclophilin a as potential molecular targets in non-small cell lung cancer. Cancer Res 63:1652–1656

Cannistra SA (2004) Cancer of the ovary. N Engl J Med 351:2519–2529

Chan EC, Koh PK, Mal M et al (2009) Metabolic profiling of human colorectal cancer using high-resolution magic angle spinning nuclear magnetic resonance (HR-MAS NMR) spectroscopy and gas chromatography mass spectrometry (GC/MS). J Proteome Res 8:352–361. doi:10.1021/pr8006232

Choi S, Kim J, Cho K et al (2010) Sequential Fe3O4/TiO2 enrichment for phosphopeptide analysis by liquid chromatography/tandem mass spectrometry. Rapid Commun Mass Spectrom 24:1467–1474. doi:10.1002/rcm.4541

Chung L, Moore K, Phillips L et al (2014) Novel serum protein biomarker panel revealed by mass spectrometry and its prognostic value in breast cancer. Breast Cancer Res 16:R63. doi:10.1186/bcr3676

Colangelo CM, Chung L, Bruce C et al (2013) Review of software tools for design and analysis of large scale MRM proteomic datasets. Methods 61:287–298. doi:10.1016/j.ymeth.2013.05.004

Coley CM, Barry MJ, Fleming C et al (1997) Early detection of prostate cancer. Part I: prior probability and effectiveness of tests. The American College of Physicians. Ann Intern Med 126:394–406. doi:10.7326/0003-4819-126-5-199703010-00010

Cornett DS, Reyzer ML, Chaurand P et al (2007) MALDI imaging mass spectrometry: molecular snapshots of biochemical systems. Nat Methods 4:828–833. doi:10.1016/j.jprot.2012.06.015

DeSouza L, Diehl G, Rodrigues MJ et al (2005) Search for cancer markers from endometrial tissues using differentially labeled tags iTRAQ and cICAT with multidimensional liquid chromatography and tandem mass spectrometry. J Proteome Res 4(2):377–386. doi:10.1021/pr049821j

Deutsch EW, Lam H, Aebersold R (2008) Data analysis and bioinformatics tools for tandem mass spectrometry in proteomics. Physiol Genomics 33:18–25. doi:10.1152/physiolgenomics.00298.2007

Domon B, Aebersold R (2006) Mass spectrometry and protein analysis. Science 312:212–217. doi:10.1126/science.1124619

Ehmann M, Felix K, Hartmann D et al (2007) Identification of potential markers for the detection of pancreatic cancer through comparative serum protein expression profiling. Pancreas 34:205–214

Felitsyn N, Peschke M, Kebarle P (2002) Origin and number of charges observed on multiply-protonated native proteins produced by ESI. Int J Mass Spectrom 219:39–62. doi:10.1016/S1387-3806(02)00588-2

Fenn JB, Mann M, Meng CK et al (1989) Electrospray ionization for mass spectrometry of large biomolecules. Science 246:64–71. doi:10.1126/science.2675315

Findeisen P, Zapatka M, Peccerella T et al (2009) Serum amyloid A as a prognostic marker in melanoma identified by proteomic profiling. J Clin Oncol 27:2199–2208. doi:10.1200/JCO.2008

Fragnoud R, Yugueros-Marcos J, Pachot A et al (2012) Isotope coded protein labeling analysis of plasma specimens from acute severe dengue fever patients. Proteome Sci 10(1):60. doi:10.1186/1477-5956-10-60

Füzéry AK, Levin J, Chan MM et al (2013) Translation of proteomic biomarkers into FDA approved cancer diagnostics: issues and challenges. Clin Proteomics 10:13. doi:10.1186/1559-0275-10-13

Geiger T, Madden SF, Gallagher WM et al (2012) Proteomic portrait of human breast cancer progression identifies novel prognostic markers. Cancer Res 72:2428–2439. doi:10.1158/0008-5472.CAN-11-3711

Goto T, Terada N, Inoue T et al (2014) The expression profile of phosphatidylinositol in high spatial resolution imaging mass spectrometry as a potential biomarker for prostate cancer. PLoS One 9:e90242. doi:10.1371/journal.pone.0090242

Grant JE, Bradshaw AD, Schwacke JH et al (2009) Quantification of protein expression changes in the aging left ventricle of Rattus norvegicus. J Proteome Res 9:53–66. doi:10.1021/pr900297f

Grover PK, Resnick MI (1995) Analysis of prostatic fluid: evidence for the presence of a prospective marker for prostatic cancer. Prostate 26:12–18. doi:10.1002/pros.2990260104

Gupta S, Chawla K (2013) Oncometabolomics in cancer research. Expert Rev Proteomics 10:325–336. doi:10.1586/14789450.2013.828947

Han X, Aslanian A, Yates JR 3rd (2008) Mass spectrometry for proteomics. Curr Opin Chem Biol 12:483–490. doi:10.1016/j.cbpa.2008.07.024

Hay SI, Battle KE, Pigott DM et al (2013) Global mapping of infectious disease. Philos Trans R Soc Lond B Biol Sci 368:20120250. doi:10.1098/rstb.2012.0250

He C, Li J, Wang R et al (2014) Metabolic profiles of 20(S)-protopanaxadiol in rats after oral administration using ultra-performance liquid chromatography/quadrupole time-of-flight tandem mass spectrometry. Rapid Commun Mass Spectrom 28:595–604. doi:10.1002/rcm.6813

Henry NL, Hayes DF (2012) Cancer biomarkers. Mol Oncol 6:140–146

Heo SH, Lee SJ, Ryoo HM et al (2007) Identification of putative serum glycoprotein biomarkers for human lung adenocarcinoma by multilectin affinity chromatography and LC-MS/MS. Proteomics 7:4292–4302. doi:10.1002/pmic.200700433

Hillenkamp F, Karas M (1990) Mass spectrometry of peptides and proteins by matrix-assisted UV laser desorption/ionization. Methods Enzymol 193:280–295. doi:10.1016/0076-6879(90)93420-P

Howard BA, Wang MZ, Campa MJ et al (2003) Identification and validation of a potential lung cancer serum biomarker detected by matrix-assisted laser desorption/ionization-time of flight spectra analysis. Proteomics 3:1720–1724. doi:10.1002/pmic.200300514

Hsu WY, Chen WT, Lin WD et al (2009) Analysis of urinary nucleosides as potential tumor markers in human colorectal cancer by high performance liquid chromatography/electrospray ionization tandem mass spectrometry. Clin Chim Acta 402:31–37. doi:10.1016/j.cca.2008.12.009

Hu S, Arellano M, Booentheung P et al (2008) Salivary proteomics for oral cancer biomarker discovery. Clin Cancer Res 14(19):6246–6252. doi:10.1158/1078-0432.CCR-07-5037

Huang H, Mackeen MM, Cook M et al (2012) Proteomic identification of host and parasite biomarkers in saliva from patients with uncomplicated Plasmodium falciparum malaria. Malar J 11(1):178. doi:10.1186/1475-2875-11-178

Hwang SI, Thumar J, Lundgren DH et al (2007) Direct cancer tissue proteomics: a method to identify candidate cancer biomarkers from formalin-fixed paraffin-embedded archival tissues. Oncogene 26:65–76. doi:10.1038/sj.onc.1209755

Hwang H, Bowen BP, Lefort N et al (2010) Proteomics analysis of human skeletal muscle reveals novel abnormalities in obesity and type 2 diabetes. Diabetes 59:33–42. doi:10.2337/db09-0214

Iribarne JV, Thomson BA (1976) On the evaporation of small ions from charged droplets. J Chem Phys 64:2287–2294. doi:10.1063/1.432536

Jadaliha M, Lee HJ, Pakzad M et al (2012) Quantitative proteomic analysis of human embryonic stem cell differentiation by 8-plex iTRAQ labelling. PLoS One 7:e38532. doi:10.1371/journal.pone.0038532

Jain S, Saxena S, Kumar A (2014) Epidemiology of prostate cancer in India. Meta Gene 2:596–605. doi:10.1016/j.mgene.2014.07.007

Kang SM, Sung HJ, Ahn JM et al (2011) The Haptoglobin β chain as a supportive biomarker for human lung cancers. Mol Biosyst 7:1167–1175. doi:10.1039/c0mb00242a

Kassa FA, Shio MT, Bellemare MJ et al (2011) New inflammation-related biomarkers during malaria infection. PLoS One 6:e26495. doi:10.1371/journal.pone.0026495

Kaur P, Rizk NM, Ibrahim S et al (2012) iTRAQ-based quantitative protein expression profiling and MRM verification of markers in type 2 diabetes. J Proteome Res 11:5527–5539. doi:10.1021/pr300798z

Keshamouni VG, Michailidis G, Grasso CS et al (2006) Differential protein expression profiling by iTRAQ-2DLC-MS/MS of lung cancer cells undergoing epithelial-mesenchymal transition reveals a migratory/invasive phenotype. J Proteome Res 5:1143–1154. doi:10.1021/pr050455t

Keshishian H, Addona T, Burgess M et al (2009) Quantification of cardiovascular biomarkers in patient plasma by targeted mass spectrometry and stable isotope dilution. Mol Cell Proteomics 8(10):2339–2349. doi:10.1074/mcp.M900140-MCP200

Kruh-Garcia NA, Wolfe LM, Chaisson LH et al (2014) Detection of mycobacterium tuberculosis peptides in the exosomes of patients with active and latent M. tuberculosis infection using MRM-MS. PLoS One 9:e103811. doi:10.1371/journal.pone.0103811

Kulasingam V, Diamandis EP (2007) Proteomics analysis of conditioned media from three breast cancer cell lines: a mine for biomarkers and therapeutic targets. Mol Cell Proteomics 6:1997–2011. doi:10.1074/mcp.M600465-MCP200

Kumar C, Mann M (2009) Bioinformatics analysis of mass spectrometry-based proteomics data sets. FEBS Lett 583:1703–1712. doi:10.1016/j.febslet.2009.03.035

Laiko VV, Baldwin MA, Burlingame AL (2000) Atmospheric pressure matrix-assisted laser desorption/ionization mass spectrometry. Anal Chem 72:652–657. doi:10.1021/ac990998k

Lander ES, Linton LM, Birren B et al (2001) Initial sequencing and analysis of the human genome. Nature 409:860–921. doi:10.1038/35057062

Lecuit M, Eloit M (2014) The diagnosis of infectious diseases by whole genome next generation sequencing: a new era is opening. Front Cell Infect Microbiol 4:25. doi:10.3389/fcimb.2014.00025

Leong S, Nunez AC, Lin MZ et al (2006) ITRAQ-based proteomic Pro Fi ling of breast cancer cell response to doxorubicin and TRAIL. J Proteome Res 11(7):3561–3572. doi:10.1021/pr2012335

Li RX, Zhou H, Li SJ et al (2005) Prefractionation of proteome by liquid isoelectric focusing prior to two-dimensional liquid chromatography mass spectrometric identification. J Proteome Res 4:1256–1264. doi:10.1021/pr049751g

Liu J, Jiang T, Wei L et al (2013a) The discovery and identification of a candidate proteomic biomarker of active tuberculosis. BMC Infect Dis 13:506. doi:10.1186/1471-2334-13-506

Liu J, Jiang T, Wei L et al (2013b) The discovery and identification of a candidate proteomic biomarker of active tuberculosis. BMC Infect Dis 13(1):1. doi:10.1186/1471-2334-13-506

Liu Y, Chen J, Sethi A et al (2014) Glycoproteomic analysis of prostate cancer tissues by SWATH mass spectrometry discovers N-acylethanolamine acid amidase and protein tyrosine kinase 7 as signatures for tumor aggressiveness. Mol Cell Proteomics 13:1753–1768. doi:10.1074/mcp.M114.038273

Lopez MF, Mikulskis A, Kuzdzal S et al (2007) A novel, high-throughput workflow for discovery and identification of serum carrier protein-bound peptide biomarker candidates in ovarian cancer samples. Clin Chem 53:1067–1074. doi:10.1373/clinchem.2006.080721

MacLean B, Tomazela DM, Shulman N et al (2010) Skyline: an open source document editor for creating and analyzing targeted proteomics experiments. Bioinformatics 26:966–968. doi:10.1093/bioinformatics/btq054

Magiera S, Baranowska I, Kusa J et al (2013) A liquid chromatography and tandem mass spectrometry method for the determination of potential biomarkers of cardiovascular disease. J Chromatogr B Anal Technol Biomed Life Sci 919–920:20–29. doi:10.1016/j.jchromb.2012.12.015

McIntosh M, Fitzgibbon M (2009) Biomarker validation by targeted mass spectrometry. Nat Biotechnol 27:622–623

Mebazaa A, Vanpoucke G, Thomas G et al (2012) Unbiased plasma proteomics for novel diagnostic biomarkers in cardiovascular disease: identification of quiescin Q6 as a candidate biomarker of acutely decompensated heart failure. Eur Heart J 33:2317–2324. doi:10.1093/eurheartj/ehs162

Megger DA, Bracht T, Meyer HE et al (2013) Label-free quantification in clinical proteomics. Biochim Biophys Acta 1834:1581–1590. doi:10.1016/j.bbapap.2013.04.001

Menon U, Jacobs IJ (2000) Recent developments in ovarian cancer screening. Curr Opin Obstet Gynecol 12:39–42. doi:10.1074/mcp.R400006-MCP200

Mu AK, Bee PC, Lau YL et al (2014) Identification of protein markers in patients infected with Plasmodium knowlesi, Plasmodium falciparum and plasmodium vivax. Int J Mol Sci 15:19952–19961. doi:10.3390/ijms151119952

Neetu K, Ankit G, Ruchi T et al (2012) A review on mass spectrometry detectors. IRJP 3:33–42

Oldenhuis CN, Oosting SF, Gietema JA et al (2008) Prognostic versus predictive value of biomarkers in oncology. Eur J Cancer 44:946–953. doi:10.1016/j.ejca.2008.03.006

Olsen JV, Mann M (2013) Status of large-scale analysis of post-translational modifications by mass spectrometry. Mol Cell Proteomics 12:3444–3452. doi:10.1074/mcp.O113.034181

Papsidero LD, Wang MC, Valenzuela LA et al (1980) A prostate antigen in sera of prostatic cancer patients. Cancer Res 40:2428–2432

Percy AJ, Chambers AG, Yang J et al (2013) Multiplexed MRM-based quantitation of candidate cancer biomarker proteins in undepleted and non-enriched human plasma. Proteomics 13:2202–2215. doi:10.1002/pmic.201200316

Pernemalm M, Lewensohn R, Lehtiö J (2009) Affinity prefractionation for MS-based plasma proteomics. Proteomics 9:1420–1427. doi:10.1002/pmic.200800377

Pitt JJ (2009) Principles and applications of liquid chromatography-mass spectrometry in clinical biochemistry. Clin Biochem Rev 30:19–34

Planque C, Kulasingam V, Smith CR (2009) Identification of five candidate lung cancer biomarkers by proteomics analysis of conditioned media of four lung cancer cell lines. Mol Cell Proteomics 8:2746–2758. doi:10.1074/mcp.M900134-MCP200

Polanski M, Anderson NL (2007) A list of candidate cancer biomarkers for targeted proteomics. Biomark Insights 1:01–48

Profumo A, Mangerini R, Rubagotti A et al (2013) Complement C3f serum levels may predict breast cancer risk in women with gross cystic disease of the breast. J Proteomics 85:44–52. doi:10.1016/j.jprot.2013.04.029

Puttamallesh VN, Sreenivasamurthy SK, Singh PK et al (2013) Proteomic profiling of serum samples from chikungunya-infected patients provides insights into host response. Clin Proteomics 10(1):14. doi:10.1186/1559-0275-10-14

Raemdonck GAAV, Tjalma WAA, Coen VP et al (2014) Identification of protein biomarkers for cervical cancer using human cervicovaginal fluid. PLoS One 9(9):e106488. doi:10.1371/journal.pone.0106488

Raijmakers R, Kraiczek K, de Jong AP et al (2010) Exploring the human leukocyte phosphoproteome using a microfluidic reversed-phase-TiO2-reversed-phase high-performance liquid chromatography phosphochip coupled to a quadrupole time-of-flight mass spectrometer. Anal Chem 82:824–832. doi:10.1021/ac901764g

Ralhan R, Desouza LV, Matta A et al (2008) Discovery and verification of head-and-neck cancer biomarkers by differential protein expression analysis using iTRAQ labeling, multidimensional liquid chromatography, and tandem mass spectrometry. Mol Cell Proteomics 7(6):1162–1173. doi:10.1074/mcp.M700500-MCP200

Rao PV, Reddy AP, Lu X, Dasari S et al (2009) Proteomic identification of salivary biomarkers of type-2 diabetes. J Proteome Res 8:239–245. doi:10.1021/pr8003776

Reddy C, Chilla D, Boltax J (2011) Lung cancer screening: a review of available data and current guidelines. Hosp Pract (1995) 39:107–112. doi:10.3810/hp.2011.10.929

Reddy PJ, Ray S, Sathe GJ et al (2015a) A comprehensive proteomic analysis of totarol induced alterations in Bacillus subtilis by multipronged quantitative proteomics. J Proteomics 114:247–262. doi:10.1016/j.jprot.2014.10.025

Reddy PJ, Ray S, Sathe GJ et al (2015b) Proteomics analyses of bacillus subtilis after treatment with plumbagin, a plant-derived naphthoquinone. OMICS 19:12–23. doi:10.1089/omi.2014.0099

Rehman I, Evans CA, Glen A et al (2012) ITRAQ identification of candidate serum biomarkers associated with metastatic progression of human prostate cancer. PLoS One 7(2):1–10. doi:10.1371/journal.pone.0030885

Rydosz A (2014) Micropreconcentrator in LTCC technology with mass spectrometry for the detection of acetone in healthy and type-1 diabetes mellitus patient breath. Metabolites 4:921–931. doi:10.3390/metabo4040921

Sawyers CL (2008) The cancer biomarker problem. Nature 452:548–552

Schiess R, Wollscheid B, Aebersold R (2009) Targeted proteomic strategy for clinical biomarker discovery. Mol Oncol 3:33–44. doi:10.1038/nbt0709-622

Schwamborn K (2012) Imaging mass spectrometry in biomarker discovery and validation. J Proteomics 75:4990–4998

Shah NP, Nicoll JM, Bleickardt E et al (2006) Potent transient inhibition of BCR-ABL by dasatinib Leads to complete cytogenetic remissions in patients with chronic myeloid leukemia: implications for patient management and drug development. Blood 108: Abstract 2166.

Shi T, Gao Y, Quek SI et al (2014) A highly sensitive targeted mass spectrometric assay for quantification of AGR2 protein in human urine and serum. J Proteome Res 13:875–882. doi:10.1021/pr400912c

Suhre K, Meisinger C, Döring A et al (2010) Metabolic footprint of diabetes: a multiplatform metabolomics study in an epidemiological setting. PLoS One 5:e13953. doi:10.1371/journal.pone.0013953

Takakura M, Yokomizo A, Tanaka Y et al (2012) Carbonic anhydrase I as a new plasma biomarker for prostate cancer. ISRN Oncol 2012:768190. doi:10.5402/2012/768190

Têtu B, Brisson J (1994) Prognostic significance of HER-2/neuoncoprotein expression in node-positive breast cancer. The influence of the pattern of immunostaining and adjuvant therapy. Cancer 73:2359–2365

Tonack S, Jenkinson C, Cox T et al (2013) iTRAQ reveals candidate pancreatic cancer serum biomarkers: influence of obstructive jaundice on their performance. Br J Cancer 108(9):1846–1853, http://dx.doi.org/10.1038/bjc.2013.150

Torok Z, Peto T, Csosz E et al (2013) Tear fluid proteomics multimarkers for diabetic retinopathy screening. BMC Ophthalmol 13(1):1. doi:10.1186/1471-2415-13-40

Tsuchida S, Satoh M, Kawashima Y et al (2013) Application of quantitative proteomic analysis using tandem mass tags for discovery and identification of novel biomarkers in periodontal disease. Proteomics 13:2339–2350. doi:10.1002/pmic.201200510

van den Broek I, Sparidans RW, Schellens JH et al (2010) Quantitative assay for six potential breast cancer biomarker peptides in human serum by liquid chromatography coupled to tandem mass spectrometry. J Chromatogr B Anal Technol Biomed Life Sci 878:590–602. doi:10.1016/j.jchromb.2010.01.011

Venter JC, Adams MD, Myers EW et al (2001) The sequence of the human genome. Science 291:1304–1351. doi:10.1126/science.1058040

Villanueva J, Shaffer DR, Philip J et al (2006) Differential exoprotease activities confer tumor-specific serum peptidome patterns. J Clin Invest 116:271–284. doi:10.1172/JCI26022

Wang YF, Fu J (2014) Rapid laboratory diagnosis for respiratory infectious diseases by using MALDI-TOF mass spectrometry. J Thorac Dis 6:507–511. doi:10.3978/j.issn.2072-1439

Wang D, Lu YQ, Liu YF (2012) Identification of new markers for childhood acute lymphoblastic leukemia by MALDI-TOF-MS. Zhongguo Shi Yan Xue Ye Xue Za Zhi 20:1365–1369

Wasinger VC, Zeng M, Yau Y (2013) Current status and advances in quantitative proteomic mass spectrometry. Int J Proteomics 2013:180605. doi:10.1155/2013/180605

Witze ES, Old WM, Resing KA et al (2007) Mapping protein post-translational modifications with mass spectrometry. Nat Methods 4:798–806. doi:10.1038/nmeth1100

Woo HM, Kim KM, Choi MH et al (2009) Mass spectrometry based metabolomic approaches in urinary biomarker study of women's cancers. Clin Chim Acta 400:63–69. doi:10.1016/j.cca.2008.10.014

Xiao JF, Varghese RS, Zhou B et al (2012) LC-MS based serum metabolomics for identification of hepatocellular carcinoma biomarkers in Egyptian cohort. J Proteome Res 11:5914–5923. doi:10.1021/pr300673x

Yan LR, Wang DX, Liu H et al (2014) A pro-atherogenic HDL profile in coronary heart disease patients: an iTRAQ labelling-based proteomic approach. PLoS One 9:e98368. doi:10.1371/journal.pone.0098368

Yang MH, Tyan YC, Jong SB et al (2007) Identification of human hepatocellular carcinoma-related proteins by proteomic approaches. Anal Bioanal Chem 388:637–643. doi:10.1007/s00216-007-1263-6)

Yang J, Yang J, Gao Y et al (2014) Identification of potential serum proteomic biomarkers for clear cell renal cell carcinoma. PLoS One 9:e111364. doi:10.1371/journal.pone.0111364

Ye B, Cramer DW, Skates SJ et al (2003) Haptoglobin-alpha subunit as potential serum biomarker in ovarian cancer: identification and characterization using proteomic profiling and mass spectrometry. Clin Cancer Res 9:2904–2911

Zeng X, Hood BL, Sun M et al (2010) Lung cancer serum biomarker discovery using glycoprotein capture and liquid chromatography mass spectrometry. J Proteome Res 9:6440–6449. doi:10.1021/pr100696n

Zeng X, Hood BL, Zhao T et al (2011) Lung cancer serum biomarker discovery using label-free liquid chromatography-tandem mass spectrometry. J Thorac Oncol 6:725–734. doi:10.1097/JTO.0b013e31820c312e

Zeng J, Yin P, Tan Y et al (2014) Metabolomics study of hepatocellular carcinoma: discovery and validation of serum potential biomarkers by using capillary electrophoresis-mass spectrometry. J Proteome Res 13:3420–3431. doi:10.1021/pr500390y

Zhang D, Holmes WF, Wu S et al (2000) Retinoids and ovarian cancer. J Cell Physiol 185:1–20. doi:10.1002/1097-4652(200010)185

Zhang Z, Bast RC Jr, Yu Y et al (2004) Three biomarkers identified from serum proteomic analysis for the detection of early stage ovarian cancer. Cancer Res 64:5882–5890. doi:10.1158/0008-5472.CAN-04-0746

Zhang X, Xu L, Shen J et al (2013) Metabolic signatures of esophageal cancer: NMR-based metabolomics and UHPLC-based focused metabolomics of blood serum. Biochim Biophys Acta 1832:1207–1216. doi:10.1016/j.bbadis.2013.03.009

Zhang Y, Jiao J, Yang P et al (2014) Mass spectrometry-based N-glycoproteomics for cancer biomarker discovery. Clin Proteomics 11:18. doi:10.1186/1559-0275-11-18

Zheng H, Li R, Zhang J et al (2014) Salivary biomarkers indicate obstructive sleep apnea patients with cardiovascular diseases. Sci Rep 14(4):7046. doi:10.1038/srep07046

Zhou J (2015) Early diagnosis of pulmonary tuberculosis using serum biomarkers. Proteomics 15:6–7. doi:10.1002/pmic.201400532

Angel Philanthropy and Crowdfunding to Accelerate Cancer Research in Developing World

5

Sandipan Ray and Vural Özdemir

Abstract

Despite the continuing threats of the global noncommunicable diseases (NCDs) epidemic, the funding agenda for disruptive innovation remains pale, particularly for cancer research and development. Chief among the NCDs, cancer stands in dire straits. Over the last decade, the global cancer burden has shifted from the developed world to the less developed countries. In developing countries, 56 % of cancer cases and 64 % of cancer deaths occurred. Worrisomely, cancer survival rates are much lower in developing countries due to the difficulties in early diagnosis, first clinical presentation as advanced disease, and, importantly, lack of disruptive innovation for new diagnostics and therapeutics. Several promising cancer research endeavors are merely shelved due to the paucity of adequate long-term or rapid response financial aid to developing world cancer researchers. Alternative sources of research funding, too, are limited in less developed countries, causing a "double jeopardy" in cancer research funding. Despite their vast population sizes in the order of billions, contribution of the philanthropic sector for cancer research is very infrequent in India and the mainland China. Disruptive innovation in cancer research will not come to fruition unless we are willing to innovate "upstream" first, at the level of research funding. Angel philanthropy tends to fund research with higher risks for failure but with greater odds for disruptive innovation. We propose that the local angel

S. Ray
Department of Biosciences and Bioengineering, Indian Institute of Technology, Bombay, Powai, Mumbai, India

Present Address: Department of Clinical Biochemistry, Metabolic Research Laboratories, Wellcome Trust-Medical Research Council Institute of Metabolic Science, Addenbrooke's Hospital, University of Cambridge, Cambridge CB2 0QQ, UK
e-mail: sandipaniitbombay@gmail.com; sr738@medschl.cam.ac.uk

V. Özdemir (✉)
Faculty of Communications & Office of the President, International Technology and Innovation Policy, Gaziantep University, Gaziantep, Turkey

School of Biotechnology, Amrita University, Amritapuri, Kerala, India

Data-Enabled Life Sciences Alliance (DELSA Global) Open Innovation Platform, Seattle, WA, USA
e-mail: vural.ozdemir@alumni.utoronto.ca

S. Srivastava (ed.), *Biomarker Discovery in the Developing World: Dissecting the Pipeline for Meeting the Challenges*, DOI 10.1007/978-81-322-2837-0_5

philanthropy networks may provide the necessary boost to create disruptive innovation ecosystems for cancer research in the developing world.

Keywords
Angel investors • Cancer research • Developing world • Philanthropic funding • Disruptive innovation • Innovation systems theory • Social studies of medicine and emerging technologies

5.1 Cancer Research in Developing World: Not Old Wine in New Bottle

Despite the continuing threats of global noncommunicable diseases (NCDs) epidemic, the funding agenda for action remains uncertain (Lancet Oncol 2011). Chief among the NCDs, cancer research stands in dire straits. In majority of the developing countries, there is a considerable divergence between the cancer research funding and cancer-related mortality rate. Over the last decade, major source of cancer mortality has shifted gradually toward the less developed and developing world from the more developed countries (Are et al. 2013). According to the 2008 GLOBOCAN estimation, 56 % of the cancer cases and 64 % of the cancer deaths occurred in low- and middle-income countries, where survival rates are also much lower, primarily due to the difficulties in early diagnosis leading to presentation with advanced disease or lack of timely, effective, and safe therapeutics (Jemal et al. 2011). Consequently, there is an urgent need for adequate funding for cancer research in the developing countries to improve cancer surveillance, early diagnosis, treatment, and prevention, not to forget for the development of the next-generation biorepositories for multidisciplinary translational cancer research (Ray et al. 2013). From a discovery point of view, cancer research in developing countries is extremely crucial as the discoveries made in developed world do not invariably "translate" to the developing world setting owing to differences in socio-epidemiological background, population diversity in lifestyle, diet, exposure to various environmental risk factors, and infectious agents (Rastogi et al. 2004). Moreover, certain types of human cancers, like cervix, uterus, stomach, liver, and oral cavity cancers, occur more frequently in developing countries than the developed western world (Kanavos 2006). Global variations in cancer epidemiology and pathogenesis support the idea for regional or local cancer research and funding for science that is context sensitive and context responsive (Lingwood et al. 2008; Özdemir et al. 2013).

In recent years, with recognition of the large tumor heterogeneity and presence of multiple subclasses within a specific type or even grade of a tumor (Bedard et al. 2013), researchers are interested to address the diversities within specific types of cancers through comprehensive multi-omics characterizations of different subtypes/malignancy grades (Network 2013). Tumors with diverse genetic and other molecular alteration indicate the vital necessity of individual patient-tailored therapies (Roychowdhury et al. 2011; Chen et al. 2012). Even though highly promising and informative, such multidisciplinary cancer research requires advanced infrastructures and substantial amount of enduring financial support. In this chapter, we explain that disruptive innovation in cancer research will not come to fruition unless we are willing to innovate "upstream" first, at the level of research funding.

5.2 Philanthropy for Cancer Research in Developing World: One Size Does Not Fit All

Cancer research is a highly dynamic enterprise demanding real-time funding in response to breakthrough discoveries. Yet many scientists find that traditional government funding streams cannot cope with a rapidly changing cancer research discovery landscape. Crowdsourcing and crowdfunding have become alternative measures for scientists who are turning to the public and citizens to fund and govern their own highly innovative research. For example, a breakthrough finding by Swedish scientists against a neuroendocrine cancer, the type of tumor that killed Apple's cofounder Steve Jobs, could only be rescued from the laboratory trenches with fresh funding from 3,846 donors in the amount of £2 million (Masters 2013; http://www.telegraph.co.uk/news/health/10120853/Is-it-time-for-the--public-to-start-funding-cancer-research.html.).

This telling example underscores the need for rapid fire funding mechanisms that can create and sustain breakthrough research in the field of cancer. It also speaks to the growing patient-led research and advocacy that are shaping both cancer research funding and agenda. Yet, this begs a further question: if governmental support for cancer research is insufficient even in well-developed countries in the west, what to do in developing world?

Philanthropic initiatives such as the Gabrielle's Angel Foundation for Cancer Research (USA and UK) and the National Cancer Coalition (NCC, USA) are providing considerable financial support for cancer research in developed world. Unfortunately, in developing countries, several promising cancer research endeavors are merely shelved due to the paucity of adequate funding, suffering from a "double jeopardy" situation, both from the governments and the philanthropy funding streams (Fernandez et al. 2012; Marko 2013).

Insofar government or traditional state research funding, developing world is not endowed with sources such as the NCI (National Cancer Institute), the AICR (Association for International Cancer Research), the AACR (American Association for Cancer Research), the ICRP (International Cancer Research Partnership), the GLOBOCAN, and the UICC (Global Cancer Control). International initiatives such as the Global Task Force on Expanded Access to Cancer Care and Control in Developing Countries (GTF.CCC), established with an intention to accelerate the expansion of cancer care and control in countries of low and middle income (Farmer et al. 2010), are certainly important, but remain insufficient to meet the demand for cancer research.

In the case of philanthropy, there has been curiously a vast paucity of support in developing countries. Lack of angel investor funding is particularly noteworthy as this particular form of philanthropy funding is more risk tolerant embracing higher risk research in return for breakthrough or disruptive innovation. Moreover, many of the angel philanthropists are former entrepreneurs who have survived the venture capital world and have interest in investing not only in personal returns but also for broader social causes that impact many citizens such as cancer research. This last point is important as not all philanthropies are identical, markedly differing in the human value systems that ultimately shape the types of research they fund and the degree of risk they are willing to tolerate.

Global funding for cancer research differs considerably between the developing and developed world (Eckhouse et al. 2008). While in the developed world a significant part of the overall cancer research funding is additionally contributed by the private sector, NGOs, angel investors, venture capital, patient advocacy groups, and charities, developing world by and large depends on the public sector (state government, central government, and government-based industries) alone. For instance, funding for cancer research in Europe is divided nearly 50:50 between philanthropic and governmental sources (Eckhouse et al. 2008); in contrast, angel investors or charity sources for supporting cancer research in mainland China or India is very rare (Cheng 2007). Contribution of the nonprofit philanthropic sector for cancer research is increasing effectively in Europe and the USA (Eckhouse and Sullivan 2006), estimated at over €500 million in Europe and €230 million in the USA in 2007. In the USA, the research funds generated through the philanthropic and angel investors are lesser compared to that of Europe, but there is very strong and sustained financial support from the National

Cancer Institute (NCI) in the USA, which is not the case for most of the developing nations (Eckhouse et al. 2008).

Moreover, there are compelling yet underappreciated reasons why attempting to fund disruptive innovation in a linear frame – seeking proof of concept before investments are made – will not work, particularly in developing world. Disruptive innovation breaks rank with established ways of discovery and product development. By definition, there is often very little data in the way of supporting an early proof of concept study while a disruptive innovation is unfolding. This is the grand challenge for both academics and industrialists: how do we cultivate disruptive innovation when there is little proof of concept?

Disruptive innovation funding is even harder in the developing world. A host of interwoven factors such as poverty, street and domestic violence, and labile physical and political milieu create a developing world innovation climate that is highly dynamic but often fragile. Linear models cannot stay well ahead of such rapidly fluctuating social dynamics in developing world to create a break from the status quo to enable disruptive innovation. In developing countries, one has to act in the face of unknown unknowns and rely on tacit instincts painstakingly cultivated from real-life field experiences in a given innovation domain.

In much the same way as riding a bicycle that cannot be easily codified in a linear manner, disruptive innovation demands a blend of nonlinear tacit knowledge on intended application contexts, some "street smarts," and embracing risks. But neither the traditional government funding streams nor the venture capital (VC) funders resonate warmly with such risk taking without linear proof of concept or are willing to rely on tacit knowledge that can foster the emergence of disruptive innovations.

5.3 Innovating Research Funding: Angel Philanthropy

The "trickle-down theory" refers to the idea that economic gains and funding to upper income levels (or richer countries) will benefit poorer members of society, by improving the economy (or the cancer research funding for our discussion in this paper), as a whole. On the other hand, given the double jeopardy of limited or scarce government and philanthropy funding, not to mention near absence of visible angel investors groups, developing world scientists are facing a growing gulf with their developed world counterparts. Yet, the future is not bleak for poor nations. There remains much untapped "citizen power" and enthusiasm, as well as angel investors who remain unorganized or not immediately visible to researchers with worthy breakthrough research findings. We suggest that rather than relying on the usual trickle-down economics and the diffusion of research funding from richer to poorer nations, developing world can move to garner more self-governing funding mechanisms by creating novel and timely philanthropy, angel investor networks in particular.

Rapid turnaround in funding decisions and acting when no one else will are crucial ingredients that cultivate disruptive innovation. Angel philanthropists fund innovative high-risk ideas at an early stage of development, and frequently without hard data on proof of concept. It is not uncommon that angel investors are themselves former entrepreneurs who had a successful financial exit in biotechnology startup ventures. Thus, they tend to act on their gut feeling and time-tested tacit hands-on field experiences to gauge the merit of a new idea rather than lengthy due diligence on paper. A personal introduction or reference by a trusted confidant might be at times all that is needed for an angel investor to move forward in funding an innovative idea. Importantly, and distinct from governments or VC funders, what motivates the angels is often noble causes and not merely a high return on their investments.

Unlike VC organizations, angel investors provide seed funding for entrepreneurs' ideas from their own personal funds, usually in amounts close to or less than US$100,000. Others, so-called super angel investors, might fund up to US$1M. Typically, an angel investor would, however, provide much less funding than a VC funder but at a greatly faster speed and at a very early stage of a promising new innovation. Because angel investors tend to fund high-risk ideas, their failure rates are higher. But they are more likely to seize the opportunity for truly disruptive innovation.

Looking forward, the generation of crowdfunded research funding by angel investors and the economies of scale brought about by the large populations in developing world could be a promising strategy to accelerate scientific research at a grassroot level in the current age of big data where harnessing the potential of vast datasets cannot be achieved by small expert communities alone (Özdemir et al. 2013). In this vein, it is noteworthy that the editorial department of Science Postprint (http://www.spp-j.com/), an open access journal in Asia, has recently initiated a crowdfunding system for promoting medical and scientific research. Even though such financial support mechanisms may not be sufficient for establishing sophisticated state-of-the-art facilities or supporting long-term multidisciplinary cancer research, they are certainly an invitation to rethink cancer research funding in the developing world beyond a passive vision to adopt citizen agency and initiative.

These efforts for philanthropy, angle investors in particular, and crowdfunding are not merely relevant for securing the funds to do cancer research. They are also a formidable means to correct for lacunas in cancer research where the end-user communities' needs and perspectives can be brought to the scientific "design table" through extended peer review enabled in the process of crowdfunding. Traditional academic research confined within the walls of academic hallways rarely pose the opportunity to engage with citizens and patients (apart from their recruitment as research subjects) where a more active role is provided for codesign and co-funding of cancer research by classical technical experts and citizens. Indeed, there is a growing recognition that the open user innovation and citizen science, in particular, can markedly accelerate design and implementation of disruptive innovation (Vayena and Prainsack 2013). To this end, citizens' "experience-based expertise" remains imperative and important more than ever for citizen science (Vayena and Prainsack 2013), and its potential applications in cancer research specifically, and NCDs more broadly, in the developing world. Time is ripe now to act on the potentials of citizen science and crowdfunding. Focusing on the nuances of different types of philanthropy and building of angel philanthropy networks together with citizens offers additional "out of the box" funding that can cultivate disruptive innovation in cancer research, both in developed and developing world.

The proposed new funding conceptual framework, named as crowdfunding 2.0 (CF 2.0 in short), is anticipated to scale up and accelerate classic angel philanthropy by combining angel investors' funding capacity with citizens' smaller amounts of contributions that rapidly accrue with the economies of scale provided by large populations such as in India (Fig. 5.1). But equally important, CF 2.0 offers a vast potential for philanthropists and citizens to work together, starting from an upstream funding and design stage of knowledge-based innovations. This would also ensure that the potential professional blind spots are also remedied in the course of funding and implementation of science for citizens to know the needs of the populations, and thus, they could usefully steer angel philanthropists to projects that can address the real-life needs of the innovation user communities such as the citizens. Such coming together of the research funders, scientists, and knowledge user communities to fund, design, and implement projects around societally significant projects is termed here as "social and responsible innovation." The salient characteristics of the CF 2.0 are summarized in Table 5.1.

5.4 Concluding Remarks

Unprecedented global health challenges such as NCDs demand unprecedented research funding solutions that bridge the local and global realities. It is not expected that philanthropic angel investment will suddenly dominate and accelerate the overall cancer research funding in developing world, but this mechanism warrants further

Fig. 5.1 A science of "hows" for next-generation crowdfunding 2.0 whereby collaboration and deliberation on research design, funding, implementation, and translation are co-created and co-managed by scientists, citizens, and angel philanthropists

Table 5.1 Crowdfunding 2.0 (CF 2.0): citizens and angel philanthropists

Characteristics
A wide range of funding amount is possible. For example, CF 2.0 offers the potential to raise microgrants for artisan creative projects or larger funds equal or more than government, industry, or angel philanthropy alone. CF 2.0 is more likely to fund high-risk ideas that government, traditional industry funding streams, or venture capital may not traditionally support
User and "social innovation" orientation
Embracing disruptive blue skies innovation
Rapid funding
Seeking "extended peer review" (beyond classic technical experts) of innovation through multiple scientific conceptual frames; intent to correct for scientists' and innovators' blind spots
Opportunity to fund junior investigators, citizen scientists with no publication record, "pathmaker" infrastructure projects, and persons with nontraditional curriculum vitae
Collective leadership that recognizes individual scholarship

consideration and analysis in the context of the developing counties.

Acknowledgements This study was supported by the Department of Biotechnology, India grant (No. BT/PR14359/MED/30/916/2010 and BT/PR13562/MED/12/451/2010) to SS.

Competing Interests The authors declare no competing interests.

References

[No authors listed] (2011) Two days in New York: reflections on the UN NCD summit. Lancet Oncol 12:981. doi: 10.1016/S1470-2045(11)70272-8

Are C, Rajaram S, Are M et al (2013) A review of global cancer burden: trends, challenges, strategies, and a role for surgeons. J Surg Oncol 107:221–226. doi:10.1002/jso.23248

Bedard PL, Hansen AR, Ratain MJ, Siu LL (2013) Tumour heterogeneity in the clinic. Nature 501:355–364. doi:10.1038/nature12627

Chen R, Mias GI, Li-Pook-Than J et al (2012) Personal omics profiling reveals dynamic molecular and medical phenotypes. Cell 148:1293–1307. doi:10.1016/j.cell.2012.02.009

Cheng MH (2007) Cancer research funding in Asia. Mol Oncol 1:135–137

Eckhouse S, Sullivan R (2006) A survey of public funding of cancer research in the European union. PLoS Med 3:e267

Eckhouse S, Lewison G, Sullivan R (2008) Trends in the global funding and activity of cancer research. Mol Oncol 2:20–32. doi:10.1016/j.molonc.2008.03.007

Farmer P, Frenk J, Knaul FM et al (2010) Expansion of cancer care and control in countries of low and middle income: a call to action. Lancet 376:1186–1193. doi:10.1016/S0140-6736(10)61152-X

Fernandez JM, Stein RM, Lo AW (2012) Commercializing biomedical research through securitization techniques. Nat Biotechnol 30:964–975

Jemal A, Bray F, Center MM et al (2011) Global cancer statistics. CA Cancer J Clin 61:69–90. doi:10.3322/caac.20107

Kanavos P (2006) The rising burden of cancer in the developing world. Ann Oncol 17(Suppl 8):viii15–viii23

Lingwood RJ, Boyle P, Milburn A et al (2008) The challenge of cancer control in Africa. Nat Rev Cancer 8:398–403. doi:10.1038/nrc2372

Marko NF (2013) The cancer megafund: determinants of success. Nat Biotechnol 31:492–494. doi:10.1038/nbt.2603

Masters (2013) Is it time for the public to start funding cancer research? The telegraph. Available from http://www.telegraph.co.uk/news/health/10120853/Is-it-time-for-the-public-to-start-funding-cancerresearch.html

Network TC (2013) Corrigendum: comprehensive genomic characterization defines human glioblastoma genes and core pathways. Nature 494:506. doi:10.1038/nature11903

Özdemir V, Badr KF, Dove ES et al (2013) Crowd-funded micro-grants for genomics and "big data": an actionable idea connecting small (artisan) science, infrastructure science, and citizen philanthropy. OMICS 17:161–172. doi:10.1089/omi.2013.0034

Rastogi T, Hildesheim A, Sinha R (2004) Opportunities for cancer epidemiology in developing countries. Nat Rev Cancer 4:909–917

Ray S, Moiyadi A, Srivastava S (2013) Epidemiology: biorepositories for cancer research in developing countries. Nat Rev Clin Oncol 10:434–436. doi:10.1038/nrclinonc.2013.119

Roychowdhury S, Lyer MK, Robinson DR et al (2011) Personalized oncology through integrative high-throughput sequencing: a pilot study. Sci Transl Med 3:111ra121. doi:10.1126/scitranslmed.3003161

Vayena E, Prainsack B (2013) The challenge of personal genomics in Germany. Nat Biotechnol 31:16–17. doi:10.1038/nbt.2469

Delivering on the Promise of Bioeconomy in the Developing World: Link It with Social Innovation and Education

6

Sandipan Ray, Sanjeeva Srivastava, Shyam Diwakar, Bipin Nair, and Vural Özdemir

Abstract

In developing countries where numerous factors such as rapid population growth and entrenched social problems hinder equitable economic growth and education, research and development (R&D) are often neglected as well. But the importance of R&D extends beyond science. The capacity to generate and advance their scientific scholarship is important for all countries – for such independent scientific thinking skills might also empower the citizens' capacity and will to think democratically in a global interdependent world. Social innovation is explained here as a form of responsible innovation that brings together funders, scientists, and knowledge user communities to address long-standing and/or entrenched societal problems. Moreover, in social innovation, the user communities such as citi-

S. Ray (✉)
Department of Biosciences and Bioengineering, Indian Institute of Technology, Bombay, Powai, Mumbai, India

Present Address: Department of Clinical Biochemistry, Metabolic Research Laboratories, Wellcome Trust-Medical Research Council Institute of Metabolic Science, Addenbrooke's Hospital, University of Cambridge, Cambridge CB2 0QQ, UK
e-mail: sandipaniitbombay@gmail.com; sr738@mcdschl.cam.ac.uk

S. Srivastava
Department of Biosciences and Bioengineering, Indian Institute of Technology Bombay, Powai, Mumbai 400076, Maharashtra, India

S. Diwakar • B. Nair
Amrita School of Biotechnology, Amrita Vishwa Vidyapeetham, Amritapuri, Kollam, Kerala, India

V. Özdemir (✉)
Faculty of Communications & Office of the President, International Technology and Innovation Policy, Gaziantep University, Gaziantep, Turkey

School of Biotechnology, Amrita University, Amritapuri, Kerala, India

Data-Enabled Life Sciences Alliance (DELSA Global) Open Innovation Platform, Seattle, WA, USA
e-mail: vural.ozdemir@alumni.utoronto.ca

S. Srivastava (ed.), *Biomarker Discovery in the Developing World: Dissecting the Pipeline for Meeting the Challenges*, DOI 10.1007/978-81-322-2837-0_6

zens can also contribute to the scientific design and funding beyond a passive role to merely adopt innovations developed by scientific experts. The overall success of developing nations thus rests on building successful linkages of the education ecosystem with social innovation and bioeconomy. To this end, E-learning endeavors and the virtual biotechnology labs are novel initiatives that are rapidly transforming society in the developing world. Distance education and E-learning and open learning endeavors are certainly advantageous for the resource-limited developing countries, where the numbers of potential learners are much higher than the number of well-experienced teachers and educational institutes capable of providing the required infrastructures for basic and advanced scientific education. India, in particular, has had strikingly innovative and forward-looking investments in biotechnology distributed learning practices that can illuminate the global society of scientists and citizens. In this chapter we will highlight the fundamental need and present scenario of virtual laboratories in advanced sophisticated life science education in the developing world.

Keywords
Awareness • Bioeconomy • Developing world • E-learning • Virtual labs

6.1 Leapfrogging Developing World: Rise of Virtual Biotechnology Labs and E-Learning Platforms

Many universities, governments, and supranational bodies in developed world have been actively investing in the biotechnology education, with an eye to cultivate knowledge society and bioeconomy (Fig. 6.1). In the European Union, the Europe 2020 economic reform and growth agenda is firmly focused on knowledge-based innovation (European Commission, Europe 2011). In the US National Science Foundation, the Office of Cyberinfrastructure has recently highlighted the emerging role of distributed learning and Campus 2.0 initiatives (National Science Foundation, Advisory Committee for Cyberinfrastructure (ACCI) Task Force Reports 2014; Waldrop 2013). In 2010 in the USA, the then Secretary of State Hillary Rodham Clinton has endorsed the vision "civilian power" for distributed and broad citizen engagement for global development and diplomacy for peace and prosperity (Clinton 2010).

The importance of research and development (R&D) extends beyond science. The capacity to generate and advance their scientific scholarship is important for all countries – for such independent scientific thinking skills might also empower the citizens' capacity and will to think democratically in a global interdependent world. For the developing world and resource-limited countries, the biotechnology and life sciences education is deemed a centerpiece in the hopes for a better life (Reddy et al. 2011). Yet the road ahead in life sciences education is not straightforward for many educators and development strategists in resource-limited countries.

Consider, for example, a country like India with a population of 1.3 billion, representing nearly 20 % of the world population. Such population size is both a veritable opportunity for the life sciences-enabled bioeconomy and a predicament to bring the life sciences education to scale in a country of immense population and geographical distribution. Learning from the lessons in developed and rich countries, the developing world can however seize the opportunity to leapfrog ahead on the global innovation and prosperity curve.

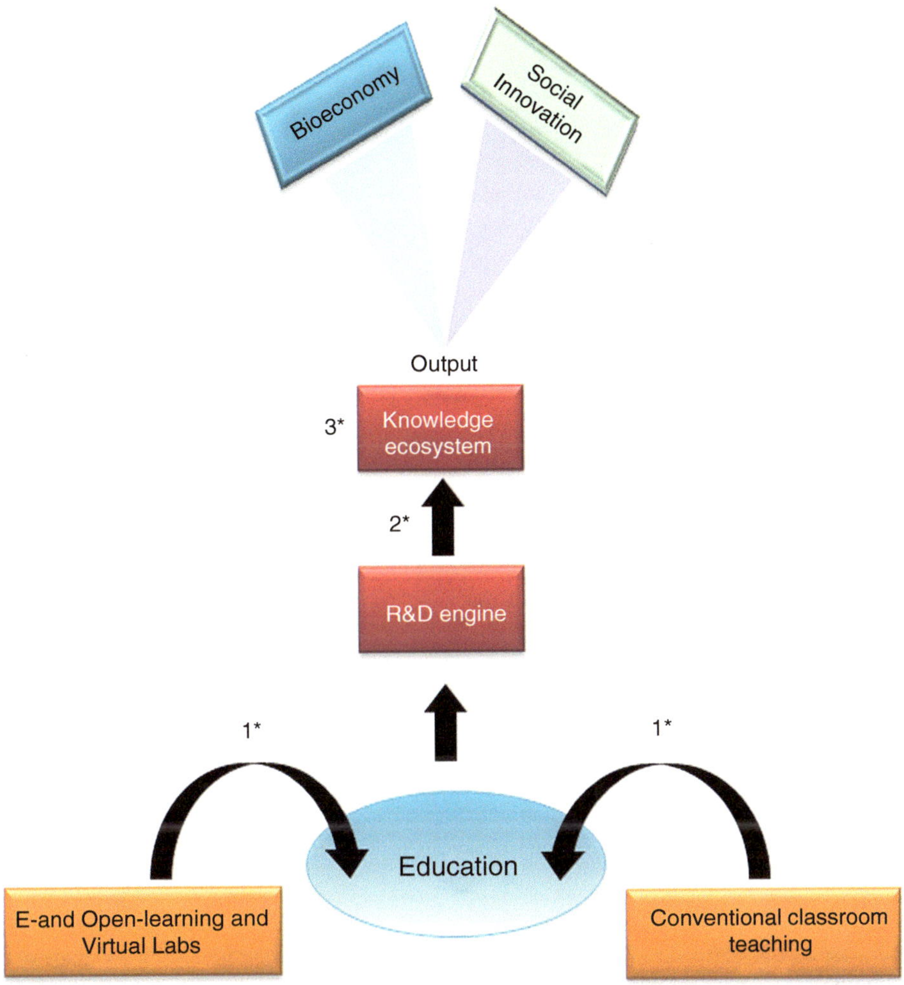

Fig. 6.1 Linkages among E-learning and open learning, biotechnology virtual labs, and traditional teaching with bioeconomy and social innovation. E-learning and open learning and virtual labs serve 1* = *enabling function* for research through life sciences graduate research education, 2* = *linkage function* for connection of R&D to knowledge ecosystem, and 3* = *diffusion function* extending the reach of new knowledge to the realm of bioeconomy and social innovation, e.g., by new market creation and value-added knowledge products

Additionally, there are opportunities to be seized that can serve the efforts to create a more equitable world through science-based initiatives. In this vein, we note the new concept of social innovation as a form of responsible innovation that brings together funders, scientists, and knowledge user communities to address long-standing and/or entrenched societal problems. Moreover, in social innovation, the user communities such as citizens can also contribute to the scientific design and funding beyond a passive role to merely adopt innovations developed by scientific experts.

We believe the developing world is well poised for social innovation and Campus 2.0 through responding to the attendant distributed biotechnology learner communities (Fig. 6.2). The key idea for leapfrogging the developing world is that they (resource-limited countries) need not be a victim but instead rapidly build on the experiences (errors and successes) of developed countries. The key to success is dependent on the ability to create a level of connectivity, a collective intelligence ecosystem where technology and human resources are interconnected in a synergistic manner, thereby fostering distance

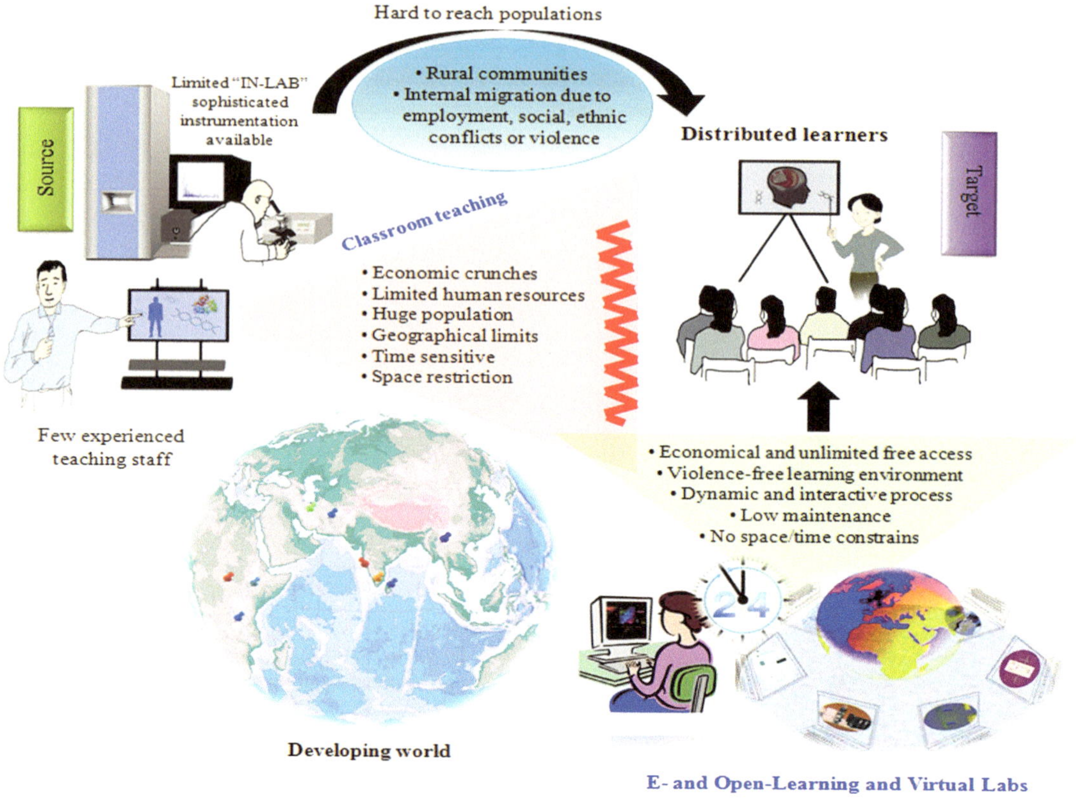

Fig. 6.2 The analysis of classroom teaching versus E-learning and open learning and virtual labs in developing world

education and open learning over spatial and temporal boundaries.

In the developing countries, the numbers of potential learners are much higher than the number of well-experienced teachers and educational institutes capable of providing quality classroom teaching. The massive open online courses (MOOCs) have risen in popularity in developed countries (Waldrop 2013), but still remain as a relatively unknown entity in the developing world as an effective platform alternative to conventional classroom teaching. Comprehensive web-based distance learning in life sciences is highly demanding due to complicated and multidimensional nature of the subject (Nilsson 2003). The success of online educational resources (OERs) depends on various essential components such as convergence for the common metadata, striking a balance between experts and community expectations for quality, community input, and interoperability (Porcello and Hsi 2013).

The "mega-universities" are newer concepts that are in the works in various locales around the world. They are built on an ethos to benefit many students particularly in developing countries and to disseminate online science education to rural communities. On the other hand, the proper teaching of laboratory-oriented scientific contents and technical training through traditional web-based online interphases have proven extremely difficult (Normile 2001). In order to circumvent these obstacles, virtual biotechnology labs, where simulators are used to create interactive tutorials that provide students a visual demonstration of techniques, are coming to the fore (Waldrop 2013). Virtual labs serve as a simulated model of a real laboratory and provide a risk-free learning environment, where students learn from repeated practices round-the-clock, at their own pace, without requiring any active involvement of teaching personnel, expensive resources, or high-maintenance equipments (Fig. 6.2). Consequently, static and remotely triggered virtual labs are among the most promising candidates in the E-learning toolbox and add a new dimension to the regular open learning courses by making them more dynamic and interactive (Huang 2004). Inclusion of virtual labs in the course curriculums of open universities in developing worlds like Universitas Terbuka (UT) of Indonesia, Open University of Sri Lanka (OUSL), and Indira Gandhi National Open University (IGNOU) of India, among others, is already reaching out to large student populations and has ambitions to inclusively embrace rural populations as a part of the efforts to create a knowledge ecosystem cultivating bioeconomy and twenty-first-century knowledge society (Ozdemir 2013).

6.2 Lessons Learned from Biotechnology E-Learning Practices in India

India is playing an increasingly significant role in development of high-quality educational curricula organized as scientific courses for global distribution (Box 6.1). The Union Minister for Human Resource Development (MHRD) and the Government of India have launched the "Sakshat Virtual Labs Project," a collection of 91 virtual laboratories containing hundreds of experiments in nine disciplines of science and engineering, as part of the MHRD's National Mission on Education through Information and Communication Technology (NMEICT) for provision of readily accessible high-quality education across the globe (Ray et al. 2012a). Additionally, a sizable selection of E-learning programs, including the National Programme on Technology Enhanced Learning (NPTEL), Open Source Courseware Animations Repository (OSCAR), Aakash E-Learning Programs, and National Knowledge Network (NKN) (Box 6.1), is at different stages of development to accelerate distance learning endeavors in educational institutes and interested learner communities. While several research institutes and universities across the country are involved actively in the present E-learning venture, the Indian Institutes of Technology (IITs) spread throughout the country and the Amrita University in south India are among the leading institutions associated with the development of web-based educational courses in different disciplines of life sciences.

To understand user perceptions, 12 hands-on workshops for 600 university students and 250 teachers were conducted at various institutions in India. A set of TAM and IEEE OER based feedback questions were provided to the virtual lab students and their individual responses were tabulated for further references. Seventy percent of students indicated virtual labs complemented classroom biotechnology courses and were appropriate as complementary teaching tool. For an evaluation, 67 % rated virtual labs served as online reference material to improve conceptual knowledge indicating usability and repeatability of such tools (Fig. 6.3). The high correlations to such acceptance were also supported by 96 % of students who indicated that their laboratory syllabus was comprehensively addressed by virtual labs. Our diffusion studies also included 10 % feedback inputs by users who reported operational and usage difficulties with virtual labs and highlighted network connectivity issues and insufficient computer training (Figs. 6.2 and 6.3).

Box 6.1 Notable Biotechnology Virtual Lab and E-Learning Initiatives from India:

- "Sakshat" Virtual Biotechnology Engineering Labs: http://www.vlab.co.in/
- Technology Enhanced Learning (NPTEL): http://nptel.iitm.ac.in/
- Open Source Courseware Animations Repository (OSCAR): http://oscar.iitb.ac.in/oscarHome.do
- Aakash E-learning Programs: http://www.aakash.ac.in/aakash-e-learning-courses
- National Knowledge Network (NKN): http://www.nkn.in/
- Amrita Virtual Interactive E-Learning World (A-VIEW): http://aview.in/

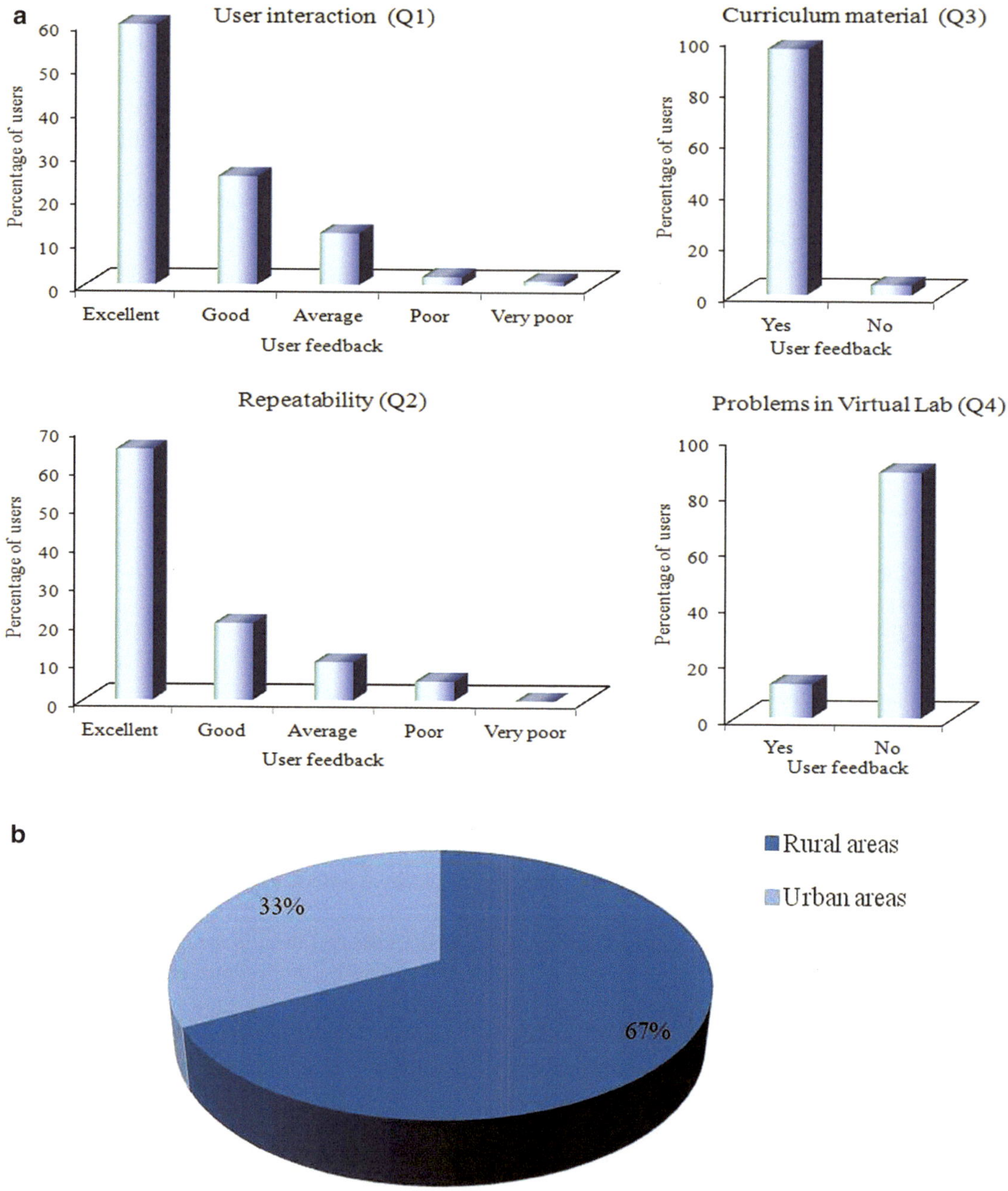

Fig. 6.3 (**a**) Analysis on user adaptability. Plots show user response (yes) percentages to easy user interaction, repeatability, curriculum material, and usage issues in virtual labs. (**b**) Usage of virtual labs: rural versus urban trends in using virtual laboratories in teaching process

6.3 A Holistic Approach to Biotechnology Teaching in Developing World

6.3.1 Science, Society, and Human Values

A look at the past twentieth century is instructive on the ways in which science has been framed around a rather narrow and deterministic technical discourse (Reddy et al. 2011; Yearley 2009). On the other hand, developing countries already do have strikingly innovative and forward-looking examples of a broader "socio-technical" teaching curricula that embed technology and scientific practices in social values, thus making science and technology teaching socially robust and sustainable. Occasionally, it is not uncommon for some life sciences teaching curricula to be embedded in spiritual practices prevalent within the large segments of the Indian society. An illustrative example of coexistence of such progressive socio-technical curricula, in a context of a rural community university specialized in E-learning initiatives, is the Amrita University in Kerala, India. The Amrita campus is completely interactive, multidisciplinary, and multimedia-enabled virtual campus, closely attuned to social values, including those of the rural society and its priorities. These include, for example, the Amrita Virtual Interactive E-Learning World (A-VIEW), Amrita-ISRO-Cognizant Technology Solutions Satellite Network for Education and Research, and the Sakshat Amrita Virtual Labs (Diwakar et al. 2011). A-VIEW is currently being used by over 350 universities and over 1,000 colleges from different parts of the country; it is also dedicated to provide free open education through collaborative synchronized sharing of 2D and 3D animations and videos with rural and remote geographically dispersed communities. At the Indian Institute of Technology Bombay (IITB), we developed the Virtual Proteomics Lab and the attendant E-learning resources for proteomics with a view to disseminate high-quality educational contents globally (Ray et al. 2012a; Srivastava et al. 2013). This effort had scaled up internationally and is now being used as a tutorial under the International Proteomics Tutorial Programme conducted by the Human Proteome Organization (HUPO) and the European Proteomics Association (EuPA) (Ray et al. 2012b). These novel E-learning resources established and presently flourishing in India serve as valuable platforms for both students and researchers in a transdisciplinary capacity across the globe.

6.3.2 Will Tools Define How Rural or Geographically Challenged Areas Improve?

There have been extensive evidences of usage of web-based learning platforms in classrooms over the last few decades (Allen and Seaman 2010; Flowers 2011). With our case studies and tests, we looked at the challenges of remote areas in India and the ability of students repeating the experiments (Swan 2003). Recent studies have shown that virtual labs have remarkable role in changing student's learning experience (Radhamani et al. 2014a). Laboratory experiments are also significant while evaluating roles of active learning in education (Nair et al. 2012; Diwakar et al. 2012). Severe limitations have been noted for improving education in developing countries (Nair et al. 2012). In a preliminary study (Diwakar et al. 2011), we found students perceived such platforms positively as a component to traditional learning process. Survey data also supported that teachers demonstrated interest to include virtual labs as a teaching tool with their classroom lectures to reduce teaching workload significantly (Diwakar et al. 2011; Nair et al. 2012). Remote areas had assessment issues and such platforms enhanced student assessment (Radhamani et al. 2014b). Studies also report adaptive learning improved learner's performance level (Diwakar et al. 2014a). Significant roles limit studies within blended classroom environments (Radhamani et al. 2014a).

Most students reported increase in perceived usefulness and a relative ease of usage (Diwakar et al. 2014b). Our initial studies with rural students also indicated enhanced outcomes in individual learning scenarios and supported the view that multidisciplinary demonstrations have a positive influence, although the current application suits blended learning that combines both face-to-face and computer-mediated instructions (Diwakar et al. 2014b). We noticed that explicit interactions connecting a user to the learning environment played a critical role toward enhancing education (Radhamani et al. 2014b). With the role of economy, problems faced by several institutions, and a variety of urban and rural users in geographically and economically challenged countries, effective supporting tools help validate teaching and enhance self-organized learning process (Diwakar et al. 2014a).

6.4 Concluding Remarks

E-learning and biotechnology virtual labs and social innovation are novel initiatives and conceptual frames that are rapidly changing the culture of life sciences education in India but also in many other resource-limited settings around the globe. They are a testament that new biotechnology and scientific findings can be taught in a globally distributed manner in the spirit of the *Campus 2.0*. Moreover, the local practices in India are telling in that science is seen, at least in some locales like the Amrita University in India's South and the IITB in Mumbai, as a social as well as a technical practice, thereby calling into long-held narrow assumptions on science as a deterministic activity. To the extent that the dynamic local science and technology are aimed to linkages with global life sciences innovation and a knowledge-driven bioeconomy, the lessons learned in India in E-learning and the virtual biotechnology labs are instructive for other developing nations to leapfrog from poverty to prosperity that is well embedded in social values and sustainable practices.

In all, the E-learning, virtual labs, and social innovation endeavors in India support the idea that biotechnology, society, and human values such as solidarity can be usefully integrated so as to create an inclusive and efficient twenty-first-century knowledge ecosystem, which is both technically modern and progressively humane.

The long-standing aspiration of E-learning in the developing world should not only concentrate on distribution of advanced scientific knowledge/higher education among the students of colleges and universities but also equally on the dissemination of life sciences education at a grassroots level among the rural communities. We anticipate that such biotechnology E-learning initiatives empowered further by the new concept of social innovation will fundamentally transform agriculture, bioindustries, and medical applications and bring together scientific experts and citizen scientists in closer proximity, thus enabling twenty-first-century knowledge society and strengthening the bioeconomy and social innovation in part driven through novel biotechnologies. Perhaps what is important is the fact that E-learning and social innovation will open up traditional academic scientific institutions to broader considerations and broader societal inputs and thus cultivate vital responsible science and innovation in a world sorely in need of solidarity and sustainable responses to the challenges of twenty-first-century science and society.

Acknowledgements The financial support from the Sakshat project of National Mission on Education through ICT, Ministry of Human Resource Development (MHRD), for the establishment of Virtual Proteomics Laboratory 10MHRD005 and Clinical Proteomics Remote Triggering Virtual Laboratory 11MHRD005 is gratefully acknowledged. We thank the Chancellor of Amrita University, Sri Mata Amritanandamayi Devi, for invaluable discussions and insights on India's E-learning initiatives and the growing role for public engagement in understanding technology future(s).

Competing Interests The authors declare no competing interests.

References

Allen IE, Seaman J (2010) Class differences: online education in the United States, Newburyport. Sloan Consortium (2010), ERIC Number: ED529952, 1–30, Available from: sloanconsortium.org

Clinton HR (2010) Foreign Aff 89:13

Diwakar S, Achuthan K, Nedungadi P et al (2011) Enhanced facilitation of biotechnology education in developing nations via virtual labs: analysis, implementation and case-studies. Int J Comput Theory Eng 3:1793–8201. doi:10.7763/ijcte.2011.v3.275, Available from: www.ijcte.org

Diwakar S, Achuthan K, Nedungadi P et al (2012) Biotechnology virtual labs: facilitating laboratory access anytime-anywhere for classroom education. In: Innovations in biotechnology. Academic Press, New York, pp 379–398. ISBN: 978-953-51-0096-6, InTech. Available from: http://www.intechopen.com/books/innovations-in-biotechnology/biotechnology-virtual-labs-facilitatinglaboratory-access-anytime-anywhere-for-classroom-education

Diwakar S, Radhamani R, Sasidharakurup H et al (2014a) Usage and diffusion of biotechnology virtual labs for enhancing University education in India's urban and rural areas. In: E-learning as a socio-cultural system: a multidimensional analysis. Information Science Reference, Hershey, pp 63–83. doi: 10.4018/978-1-4666-6154-7

Diwakar S, Parasuram H, Medini C et al (2014b) Complementing neurophysiology education for developing countries via cost-effective virtual labs: case studies and classroom scenarios. J Undergrad Neurosci Educ 12:A130–A139

European Commission, Europe (2011) EU-Wide Targets, 2020. Available from: http://ec.europa.eu/europe2020/targets/eu-targets/index_en.htm

Flowers OL (2011) Investigating the effectiveness of virtual laboratories in an undergraduate biology course. J Hum Resour Adult Learn 7:110–116, ISBN: 18172105

Huang C (2004) Virtual labs: E-learning for tomorrow. PLoS Biol 2:e157

Nair B, Krishnan R, Nizar N et al (2012) Role of ICT-enabled visualization-oriented virtual laboratories in universities for enhancing biotechnology education – VALUE initiative: case study and impacts. FormaMente 7:209–229, ISBN: 1970-7118

National Science Foundation, Advisory Committee for Cyberinfrastructure (ACCI) Task Force Reports, in 2014.

Nilsson T (2003) Virtual laboratories in the life sciences. A new blueprint for reorganizing research at the European level. EMBO Rep 4:914–916

Normile D (2001) Online science is a stretch for Asia. Science 293:1623

Özdemir V (2013) OMICS 2.0: a practice turn for 21(st) century science and society. OMICS 17:1–4. doi:10.1089/omi.2012.0111

Porcello D, Hsi S (2013) Science education. Crowdsourcing and curating online education resources. Science 341:240–241. doi:10.1126/science.1234722

Radhamani R, Sasidharakurup H, Kumar D et al (2014a) Explicit interactions by users form a critical element in virtual labs aiding enhanced education - a case study from biotechnology virtual labs. In: IEEE T4E. doi: 10.1109/T4E.2014.37

Radhamani R, Sasidharakurup H, Sujatha G et al (2014b) Virtual labs improve student's performance in a classroom. eLEOT. doi:10.1007/978-3-319-13293-8_17

Ray S, Koshy NR, Diwakar S et al (2012a) Sakshat Labs: India's virtual proteomics initiative. PLoS Biol 10:e1001353. doi:10.1371/journal.pbio.1001353

Ray S, Koshy NR, Reddy PJ, Srivastava S (2012b) Virtual Labs in proteomics: new E-learning tools. J Proteomics 75:2515–2525. doi:10.1016/j.jprot.2012.03.014

Reddy PJ, Jain R, Paik YK et al (2011) Personalized medicine in the Age of pharmacoproteomics: a close up on India and need for social science engagement for responsible innovation in post-proteomic biology. Curr Pharmacogenomics Pers Med 9:67–75

Srivastava S, Özdemir V, Ray S et al (2013) Online education: E-learning booster in developing world. Nature 501:316. doi:10.1038/501316c

Swan K (2003) Learning effectiveness online: what the research tells us. Elem Qual online Educ 13–45. doi: 10.1111/j.1467-8535.2005.00519.x. Available from: cguevara.commons.gc.cuny.edu

Waldrop MM (2013) Online learning: campus 2.0. Nature 495:160–163. doi:10.1038/495160a

Yearley S (2009) The ethical landscape: identifying the right way to think about the ethical and societal aspects of synthetic biology research and products. J R Soc Interface 6(Suppl 4):S559–S564. doi:10.1098/rsif.2009.0055

Role of Proteomics in Characterization of Biosimilar Products

7

Anurag S. Rathore, Sumit K. Singh, Neh Nupur, and Gunjan Narula

Abstract

Biomarker discovery has been one of the central foci of proteomic research globally. However, in developing countries, the road map to achieving this is not very straightforward. Biomarker post regulatory checks are often known to be sold at exorbitant prices in the markets. The ever-increasing cost of healthcare together with our improving understanding of biotech therapeutic drugs has fueled the rise of biosimilars. Biosimilars are defined as biotechnological products that have been proved to be comparable to an already approved reference product in quality, nonclinical, and clinical evaluation. Discussion and resolution of the various scientific and regulatory factors that play a role in approval of biosimilars are arguably one of the most significant events over the last decade for biotechnology. Key scientific factors include the complexity of biotech products and processes, use of complex raw materials that are not always well characterized, and our relatively limited understanding of how the numerous quality attributes that define a biotherapeutic impact the product's safety and/or efficacy in the clinic. A key step toward achieving successful development of a biosimilar is to establish analytical comparability with the innovator drug. This is necessary for the biosimilar manufacturer to avail of the significant reduction in clinical data required for achieving regulatory approval. Proteomic-based analytical tools have played an important role and have gradually emerged as a major resource for characterization of analytical information to characterize these biosimilars, thereby playing a major role in the biosimilar revolution. This chapter addresses major developments that have taken place in the use of proteomics toward development of biosimilars with a focus on progress made in the last 5 years.

A.S. Rathore (✉) • S.K. Singh • N. Nupur
G. Narula
Department of Chemical Engineering, Indian Institute of Technology, Hauz Khas, New Delhi 110016, India
e-mail: asrathore@biotechcmz.com; http://www.biotechcmz.com

S. Srivastava (ed.), *Biomarker Discovery in the Developing World: Dissecting the Pipeline for Meeting the Challenges*, DOI 10.1007/978-81-322-2837-0_7

Keywords
Biosimilars • Mass Spectrometry • Label-free techniques • Gel-based methods

7.1 Introduction

Biotherapeutics, also referred to as biotech drugs or biologics, are an emerging class of treatment modalities that are produced by harnessing the protein synthetic machinery of living cells. Hormones, cytokines, monoclonal antibodies, and fusion proteins are few examples of various classes of biologics that have been instrumental in providing effective solutions to complex diseases which the conventional pharmaceutical remedies have struggled to provide (Walsh 2014). Further, with the US and European markets for biologics presently accounting for approximately $60 billion in annual sales, biopharmaceutical companies have a significant financial motivation to pursue development of biosimilars (www.alliedmarketresearch. com biosimilars market). However, the process to go from the idea to getting marketing approval for biotech drugs is long, expensive, and arduous. It is therefore not surprising to come up with reports attributing spending on biotherapeutic-based treatments between $37,000/year (for Herceptin) and staggering $200,000/year (for Cerezyme) (Kálmán-Szekeres et al. 2012).

Present interest in biosimilars has also been fueled by the fact that patents supporting the majority of the present biotech therapeutic blockbusters have either expired recently or will be expiring in the next 5 years (Hirsch and Lyman 2014). This somewhat unique situation has offered a significant opportunity to the manufacturers to develop and commercialize biosimilars to the existing therapies in the market. However, biosimilars like other biotherapeutics are complex molecules, and proving that they are structurally and functionally similar to the reference product is nontrivial (Rathore 2009a). The various structural/functional heterogeneities can arise during the long, complex process of manufacturing the therapeutic or during the storage and/or transport of the biotherapeutic (Rathore et al. 2016). Common alterations that have been reported in literature that are believed to affect safety and efficacy of the product include, but are not limited to, aggregate formation, C- and N-terminal modifications, fragmentation, oxidation, deamidation, glycosylation, conformational changes, and disulfide bond shuffling (Rathore 2009b). Apart from these attributes, the presence of host cell proteins, residual DNA, raw material-derived impurities, and other contaminants including endotoxins in the biosimilar product can also significantly affect safety and efficacy of a biotherapeutic (Eon-Duval et al. 2012; Chirino and Mire-Sluis 2004). These affects may be manifested in the form of pharmacological or immunogenic response in the patient. These attributes are referred to as critical quality attributes (CQA) (Hirsch and Lyman 2014; Rathore 2009a).

While the above mentioned concern over assessment of quality, safety, and efficacy of a biosimilar product is valid given the limited clinical data that normally these products have prior to their approval, the growing knowledge about these products over the past decade has resulted in our confidence on basing this assessment more on establishing the analytical comparability between the biosimilar and the innovator product. The latter involves the use of a large number of orthogonal, high-resolution, analytical methods that are capable of characterizing and detecting an entire gamut of changes/modifications in the product (Berkowitz et al. 2012). Successfully demonstrating highly similar analytical profile lowers the risk of finding lack of comparability with the reference product in the clinic, and this is the basis for reduced clinical data requirement from biosimilar products and the resulting significantly lower cost of development and commercialization of biosimilar products (Hirsch and Lyman 2014; Singh et al. 2016). However, the lack of harmonized global approval pathways and absence of product-specific guidelines from regulatory agencies hinder biosimilar development (Wang and Chow 2012).

Traditional means of characterization of protein structure via tools, such as circular dichroism (CD) spectroscopy, suffer from a major limitation that they generally provide information that is derived from a sum of signals arising from analysis of the many different parts of the protein that is being analyzed. Hence, while such assessments continue to be an important part of product characterization package, the fact is that they are providing an overall average over the entire structure of the biotherapeutic. As a result, it becomes an extremely difficult task to discriminate between two large signals that only differ by a small amount, as is the case when establishing analytical comparability for a biosimilar product (Berkowitz et al. 2012).

In the above context, the improvements in proteomic instrumentation over the past 6–7 years in concomitance with the improvements in chromatographic, electrophoretic separations, mass spectrometer (MS) sensitivities, and resolutions have led to the more precise characterization of physicochemical features as well as impurities in the biosimilar preparations. In particular, mass spectrometry-based structural proteomic tools are capable of performing rapid, parallel, and large-scale characterization of proteins and thereby defining identity, quantity, structure, and function of a biotherapeutic (Chen et al. 2013). Important tools that fall under this category include mass spectrometry (Fig. 7.1) coupled with the various separation methods (chromatography, electrophoretic) for protein identification and protein complex deconvolution (Beck et al. 2015), label-free techniques for protein quantitation and analysis of interaction kinetics (Myszka and Rich 2000), and gel-based methods for impurity profiling (Levy et al. 2014).

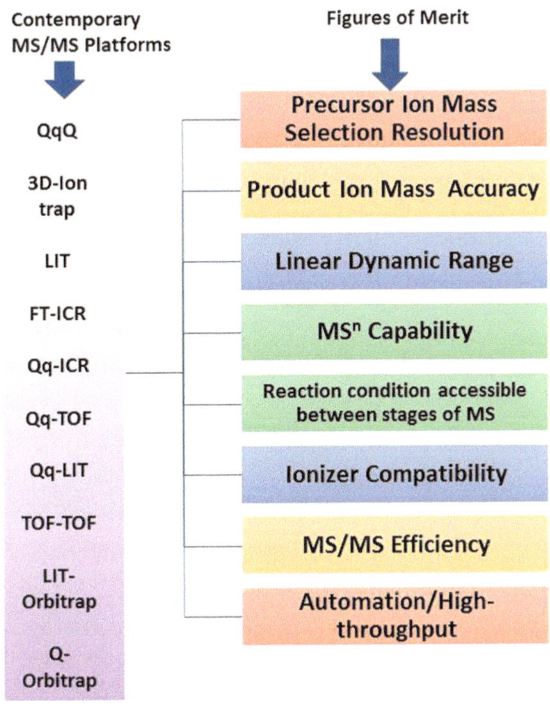

Fig. 7.1 Listing contemporary MS/MS platform with common figures of merit that enable comprehensive protein characterization. *Q* quadrupole mass filter, *q* quadrupole collision cell, *TOF* time of flight, *ICR* ion cyclotron resonance, *LIT* linear quadrupole ion trap, *FT* Fourier transform

This chapter addresses major developments that have taken place in the use of proteomics toward development of biosimilars with a focus on progress made in the last 5 years.

7.2 Structural Characterization of Biologics

Mass spectrometry gained prominence in protein characterization with the development of soft ionization techniques such as electrospray ionization (ESI) (Wilm 2011) and matrix-assisted laser desorption/ionization (MALDI) (Hillenkamp et al. 1991). These tools made mass spectrometry amenable to analysis of larger biomolecules which was hitherto limited to smaller compounds due to difficulty in transferring the former into charged gaseous phase. Though MALDI is suited for protein characterization, the associated constraints with regard to the analysis of complexes, protein flexibility, and charge states produced by the MALDI make ESI a more advantageous tool for characterization of biotherapeutics. This is quite evident from the number of publications in the scientific literature employing the latter (Srgel et al. 2010; Xie et al. 2010; Haselberg et al. 2011; Beck et al. 2012). However, significant developments have taken place in mass spectrometric analysis of proteins and peptides over time, and these have made it possible for us to characterize proteins with a single amino acid residue resolution (Singleton 2014). It will therefore not be an exaggeration to claim that mass spectrometry-based characterization has become the most highly utilized analytical technique for characterization of biotherapeutic products due to its versatility, sensitivity, and ability to decipher both small and large alterations in biologics such as sequence variants, diversity in glycosylation profiles, and other post-translational modifications (PTMs) including disulfide shuffling (Beck et al. 2012). Figure 7.2 illustrates the basic MS workflow for protein characterization. Steps having scope for further improvement are highlighted.

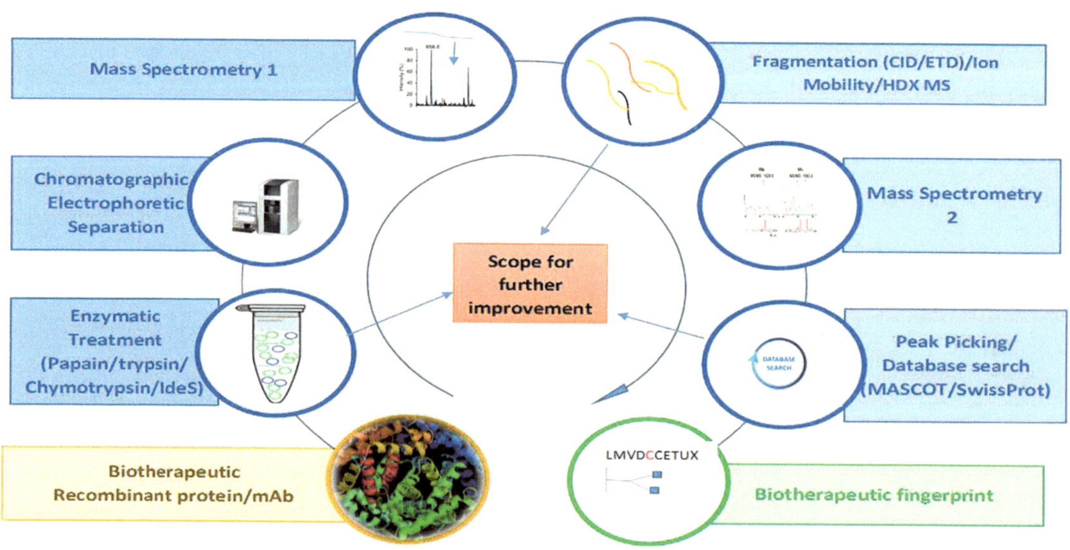

Fig. 7.2 Illustration of mass spectrometric workflow for protein identification and characterization. The characterization begins by proteolytic digestion of the therapeutic protein followed by a step of chromatographic/electrophoretic separation of the components. MS 1 analysis gives an overall picture of any possible variations/modifications in the protein structure. MS2 spectra is acquired after subjecting the peptides to fragmentation using novel functionalities attached to mass spectrometer such as electron capture dissociation/collision-induced dissociation (ECD/CID). Database search of the acquired spectrum facilitates in deriving the protein fingerprint. Inputs for the figure are obtained from the reference (Cottrell 2011)

7.2.1 Primary Sequence Assessment

One of the foremost regulatory requirements for biosimilars is to have identical amino acid sequence as the reference product even if the former differs from the latter in terms of other microheterogeneities, such as differential PTM, provided the difference does not have any clinical implications (Beck et al. 2013). In view of this, the biotech industry and other academic researchers have strived over the last two decades to create methodologies and platforms that can deliver comprehensive analytical data that can enable characterization of various facets of primary structure to establish equivalence with the innovator (Nupur et al. 2016). An ideal case is to use this structural information to quickly eliminate unsatisfying candidates at an early stage of biosimilar development.

Assessment of the primary structure begins by intact mass measurements (top-down analysis) in which accurate mass, purity, and glycosylation profile are assessed. Any increase/decrease in the mass between innovator and biosimilar indicates the presence of sequence variants or of posttranslational modifications. This analysis is typically carried out using electrospray ionization time-of-flight mass spectrometry (ESI-TOF-MS) coupled to a separation device (chromatography or electrophoresis). However, the analysis at this level can only be used to indicate the possibility of the presence of the variation without allowing us to precisely locate the type and/or nature of the variants. To achieve the latter, strategies like middle-up/middle-down/bottom-up analysis by using advanced mass spectrometry instrumentation having the capability of making measurements even in femtomole levels are supplemented with the intact mass analysis (Ayoub et al. 2013). Table 7.1 summarizes different MS-based proteomic techniques used for structural characterization of biotherapeutics. These approaches aim to detect details of even the minute variation with complete sequence coverage (Pan and Borchers 2014). Recently, an enzyme derived from *Streptococcus pyogenes* referred to as immunoglobulin-degrading enzyme of *Streptococcus pyogenes* (IdeS) has been used to cleave monoclonal antibody (mAb) therapeutics at the hinge region, thereby producing fragments that have simplified data analysis as well as the time of analysis (Beck et al. 2013, 2015; Lynaugh et al. 2013). Apart from this, advancements in the architecture of the analyzers and electronics of mass spectrometers have facilitated identification of even isotopic distribution of mAb biosimilars with a high degree of precision (Singleton 2014). With these improvements, fragmentation of intact proteins takes place directly using electron-based dissociation or by collision with neutral molecules (with electron

Table 7.1 Summary of different MS-based proteomic techniques used for structural characterization of biotherapeutics

Detailed structure Information

SL. No.	Analysis methodology	Peptide Size analyzed (In kDa)	Feature
1	Bottom Up Proteomics (BUP)	0.6-3	~99% sequence confirmation, almost complete protein fingerprint
2	Extended Bottom up proteomics (eBUP)	3-7	Near complete sequence confirmation, Nature and location of most modifications, disulphide linkages, glycosylation profiles
3	Middle down proteomics (MDP)	7-15	More complete sequence confirmation, Nature and location of more modifications
4	Top down proteomics (TDP)	>15	Confirmation of elemental composition and nature of modifications

Less structure Information

transfer dissociation/collision-induced dissociation function in mass spectrometer). These techniques are proving to be an important alternative to enzyme based proteolytic analysis and are currently used extensively for biosimilar characterization. Examples of the emerging MS formats for structural proteomics include but are not limited to native MS, ion mobility MS, and hydrogen/deuterium exchange MS (Beck et al. 2015).

7.2.2 Peptide Mapping

Peptide mapping is by far the most commonly used technique for comprehensively cataloging of any differences that may be present at the primary structure level in a biotech therapeutic. The technique is used for preliminary screening of batch-to-batch variation during bioprocess development. The constant evolution of proteomic instrumentation has advanced the technique of peptide mapping considerably. As an example, a data-independent acquisition reverse-phase peptide mapping approach has been proposed with alternate low and elevated collision energy scanning (LC/MS^E) to investigate differences in protein sequence between a biosimilar candidate and the corresponding innovator product (trastuzumab) (Xie et al. 2010). The approach allowed for identification of mutation in the heavy chain of the biosimilar. The observed difference, according to the authors, could perhaps be attributed to inconsistent integrity of expression clones which if corrected at an early stage of biosimilar development could significantly reduce the chances of product failure at later stages of product development. The uniqueness of the method adopted in this study was that the fragmentation of the precursor ions was performed independent of the prior precursor ion scan, a characteristic feature of data-dependent analysis or shotgun proteomic approach. The limitation with the latter is that the detectable dynamic range is restricted to the peptides that ionize the best. However, the proposed method allowed accurate quantification over a wide dynamic range.

7.2.3 Native Mass Spectrometry-Based Structural Characterization

Native MS is increasingly being used in probing structural biology of therapeutic drugs including complex monoclonal antibody conjugates targeting multiple molecular targets or mechanisms of action (Rosati et al. 2014). Low sample requirement with the ability to monitor dynamics and subpopulations in the same sample being probed is a characteristic feature of native MS that overcomes the limitations of traditional tools like X-ray and nuclear magnetic resonance (NMR), which require flexible parts of the protein to be removed for analysis (Lorenzen and van Duijn 2010). Since a key objective of biosimilar characterization is to determine any structural changes that impact the clinical performance of the drug, it is of prime importance that the analytical technique used allows for retention of dynamism of the molecule (functional native conformation without damaging their covalent/non-covalent interactions) while making these measurements. While ESI-TOF-MS continues to be used for intact mass measurements, use of newer versions like Orbitrap MS and others have allowed biosimilar characterization workflows with multifold enhanced resolution that have enabled identification of multiple co-occurring PTMs and other modifications. A recent publication provides a detailed and comprehensive protocol for intact mass analysis of different mAb modalities with recommendations about applications that can be accomplished using a TOF analyzer vs. others that may require higher-resolution versions of the MS (Rosati et al. 2014).

7.2.4 Ion Mobility Mass Spectrometry (IMS)-Based Structural Characterization

While it is a prerequisite to obtain optimal ionization for efficient mass analysis, the biggest concern when performing protein analysis in any MS format is to retain solution-phase structure in gas phase. Several researchers have reported occurrence of changes in protein structure upon transition from solution to ionized gas phase (Vahidi et al. 2013; Warnke et al. 2013; Konermann et al. 2014). Addition of ion mobility spectroscopy to mass spectrometry offers an attractive solution to this problem by allowing separation of ions (isomers, isobars, and conformers) that have an identical elemental composition but differ only in their structural arrangement based on differential mobility. Large molecules have higher collision cross section than small molecules thereby traversing in different times in a chamber filled with inert gas under the influence of electric field. Three common variants of ion mobility spectrometry are commonly used in protein characterization. These include drift tube (DT)-type, differential mobility analyzer (DMA)-type, and traveling wave (T-wave)-type instruments (Zhong et al. 2012). What makes ion mobility even more attractive is the possibility to couple IMS to multiple MS versions such as TOF, linear quadrupoles, trapping devices, Fourier transform ion cyclotron resonance (FTIR), and magnetic sector spectrometers, thereby harnessing the advantages of each and adapting the analysis according to the needs/questions to be answered (Harvey et al. 2011). Over the past 2 years, important applications of IMS-MS reported in literature include head-to-head comparability study of trastuzumab and cetuximab with the corresponding biosimilar and biobetter candidates (Beck et al. 2015), assessment of differences in the structure of protein reference materials and their interactions with antibodies using human growth hormone as model protein (Pritchard et al. 2013), and online monitoring of a Fab (fragment antigen binding) arm exchange and bispecific antibody formation between two humanized immunoglobulins using a T-wave-type IM-MS (Debaene et al. 2013).

7.2.5 Glycosylation Analysis

Glycosylation is well accepted as a critical quality attribute (CQA) due to its influence on the biological activity, efficacy, stability, immunogenicity, bioavailability, antibody-dependent cellular cytotoxicity (ADCC), and complement-dependent cytotoxicity (CDC) of a commercial therapeutic glycoprotein (Goochee et al. 1991; Burton and Dwek 2006). From the perspective of biosimilar development, it is not even possible to have an identical glycosylation profile as the innovator. However, the profiles need to be similar, and regulatory approval requires that the differences be justified with regard to their clinical impact. These considerations underscore the necessity of monitoring glycosylation pattern for biotherapeutics during all stages of manufacturing so as to ensure that desirable product quality, batch-to-batch consistency, and product stability have been achieved. However, this is nontrivial for the case of glycosylated proteins (Kaltashov et al. 2013), where high structural diversity and heterogeneity are often present depending on the cell line characteristics, process variables, and culture media components (Butler 2005; Velez-Suberbie et al. 2013; Kunkel et al. 2000). Complete structural elucidation of glycans requires the determination of the sugar composition, sugar sequence, monosaccharide branching, interglycosidic linkages, and anomeric configuration (Morelle and Michalski 2007). Mass spectrometry-based approaches allow for this detailed characterization of both the N-linked (glycan attached to Asn of consensus sequon Asn-x-Ser/Thr) and O-linked (glycan attached to Ser/Thr residue of the polypeptide chain) glycosylation profiles of therapeutic proteins with the required sensitivity (Narula et al. 2016).

Leveraging on the improvements in fragmentation techniques, researchers have combined collision-induced dissociation (CID), electron transfer dissociation (ETD), and CID of charge-reduced species (CRCID) derived from ETD to gain information about the sites of glycosylation modifications as well as the sequence of the peptide fragments (Wu et al. 2007). The different mechanisms attributed to CID and ETD for fragment generation provide complementary structural information and thereby enhance the confidence in assignment of structural heterogeneity.

In a similar attempt to distinguish lot-to-lot heterogeneity in N-glycosylation profile of a therapeutic (trastuzumab), researchers have used electrospray quadrupole time-of-flight mass spectrometry (ESI-qTOF-MS) to analyze global glycosylation profile at intact, peptide, and released glycan levels (Damen et al. 2009). The unique architecture of the instrument that allowed for time aligned parallel fragmentation of the mAb digest using the dual-collision cell design along with an additional dimension of ion mobility separation function helped in deciphering the glycosylation site and glycan structure in a rapid manner. Further improvements in protein characterization by mass spectrometry have been proposed with three different mass spectrometric-based approaches being used for characterization as well as quantitation of the N- and O-glycosylation profiles of glycoproteins (Morelle and Michalski 2007). The proposed approaches differ from one another in terms of the number of steps that need to be performed and the type of MS format that should be used for final analysis. The suggested metric to be followed in choosing the method is that of amount of a sample available at hand for analysis. Even more recently, researchers have aimed to identify regions of the biosimilar therapeutic that could potentially affect safety and efficacy (Chen et al. 2013). They employed a three-step LC/MS-based global proteomic approach for performing comprehensive comparability studies. The strategy included analytical characterization by peptide mapping and mass sequencing followed by targeted comparison of the different modification forms and the non-targeted comparison of the tryptic digest map.

7.2.6 Assessment of Higher-Order Structures (HOS)

Determination of protein conformation is an important component of comparability studies for biosimilars, particularly in instances where conformational changes occur without any change in primary structure/sequence. Often, these changes are triggered by initiation of degradation pathways that render the protein immunogenic by aggregate formation (Wei et al. 2014). As discussed in the previous sections of this chapter, changes/modifications/alterations at the primary level of the proteins can be detected/characterized with tandem LC/MS approaches. However, for assessing changes that result due to degradation of proteins requires probing of higher-order structures.

Pairing of hydrogen/deuterium exchange (HDX) is being adopted in the industry with high degree of success as it has been instrumental in providing detailed insights of post-translational modifications (Houde et al. 2010), conformational changes (Lodowski et al. 2010), and more importantly in comparability studies of biosimilars and innovators (Houde et al. 2011; Visser et al. 2013). The technique is based on the differential uptake of deuterium (heavier) in exchange of protons/hydrogen (lighter) by different conformations of a biomolecule (Konermann et al. 2011). The differential exchange rate corresponding to the different conformation is captured by mass spectrometry as deuterium has a higher mass than protons/hydrogen which gets reflected on the acquired mass spectrum. Apart from this, HDX-MS is also used for epitope mapping of protein therapeutics (Wei et al. 2014). Notable applications of HDX-MS particularly for biosimilar development include examining the effects of the different types of glycosylation on mAb conformation and protein receptor binding (Houde et al. 2010), conformational characterization of the charge variants of a human IgG1 monoclonal antibody (Tang et al. 2013), and developing correlation between stress-induced chemical modification and the conformation of protein therapeutics (Zhang et al. 2011). The application of HDX-MS is continuing to be adopted in more areas of biotherapeutic analysis.

7.3 Assessment of Host Cell Proteins

Host cell proteins (HCPs) and host cell DNA (HCD) are produced by the microbial or mammalian cells that are being used for expressing the product of interest. Most of these impurities are removed in the early stages of separation and their presence in the final product occurs at very low levels (<100 ppm for HCP and <10 ng per dose for HCD). While HCD is quiet easy to detect and remove due to the vastly different physicochemical properties that they have in comparison to a protein product, adequate removal of HCP is nontrivial as they represent not just a single specie but rather a population of species, some of which are bound to have very similar physicochemical properties than the biotherapeutic itself. Till recently, enzyme-linked immunosorbent assay (ELISA)-based methods were predominantly used by industry for HCP quantitation in the various process intermediates and the final product (Champion et al. 2005). However, a key limitation of ELISA is its inability to detect weakly immunoreactive proteins and to define identity of the HCP. This has fueled a lot of efforts toward finding better analytical approaches for HCP profiling (Hogwood et al. 2014).

Two techniques that have emerged as the most promising for qualitative as well as quantitative analysis of HCP are two-dimensional difference gel electrophoresis (2D-DIGE) and liquid chromatography mass spectrometry (LC/MS) with the latter involving characterization and quantification of peptides obtained from the proteolytic digestion of protein (Rathore and Bhambure 2014). 2D-DIGE has enabled improvement in quantification of HCP using 2D-PAGE (Jin et al. 2010). Also, to overcome the detection of less abundant protein spots on a 1D/2D gel, different staining agents such as SYPRO ruby protein stain against the conventional silver/coomassie stain commonly used in mass spectrometry workflows that improve the sensitivity of detection have been evaluated (Lopez et al. 2000; Hogwood et al. 2014). Among the non-gel-based methods, mass spectrometry is ubiquitously being employed for evaluating HCP. MS was used for the first time for detailed HCP characterization of a biotech drug substance/product in a recent publication. (Schenauer et al. 2012). In another publication, researchers employed mass spectrometry for quantifying residual HCP level in biotherapeutic product. The interesting aspect of this study was to derive a correlation between effects of process change on residual HCP level and basing this knowledge toward implementation of QbD and demonstrating comparability of HCP levels in biosimilars following manufacturing changes (Schenauer et al. 2013). Prior to this study, with a view to gain an understanding of how different bioprocessing steps influence HCP profiles, supernatants of mAb-producing CHO cells were analyzed using 2D-PAGE and LC/MS/MS technique (Tait et al. 2012). This study highlighted a positive correlation between HCP profile and cell viability. Previously, researchers had employed 2D liquid chromatography coupled with high-resolution mass spectrometry (2D-LC/MS) followed by high-throughput HCP quantification by liquid chromatography using 1D-LC multiple reaction monitoring (MRM) (Doneanu et al. 2012). The sensitivity, specificity, multiplexing capability, and precision of MRM coupling with MS allowed identification and quantification of HCP with sufficiently high throughput. Capillary electrophoresis-mass spectrometry has also received great interest among researchers for HCP analysis in biotech products (Zhu et al. 2016). Figure 7.3 provides an overview of the analytical techniques employed for HCP profiling.

Taking together the outcome of all these extensive HCP profiling studies, it can be safely concluded that apart from the impact that HCPs have on safety and efficacy, well-characterized HCP profiles can aid in devising efficient product development and subsequent purification strategies that in turn could boost the overall bioprocess productivity. While major advancements have been made in our ability to measure and characterize HCP, we feel that there are more needs to be done to minimize the possible risk of HCP contamination in a biotherapeutic.

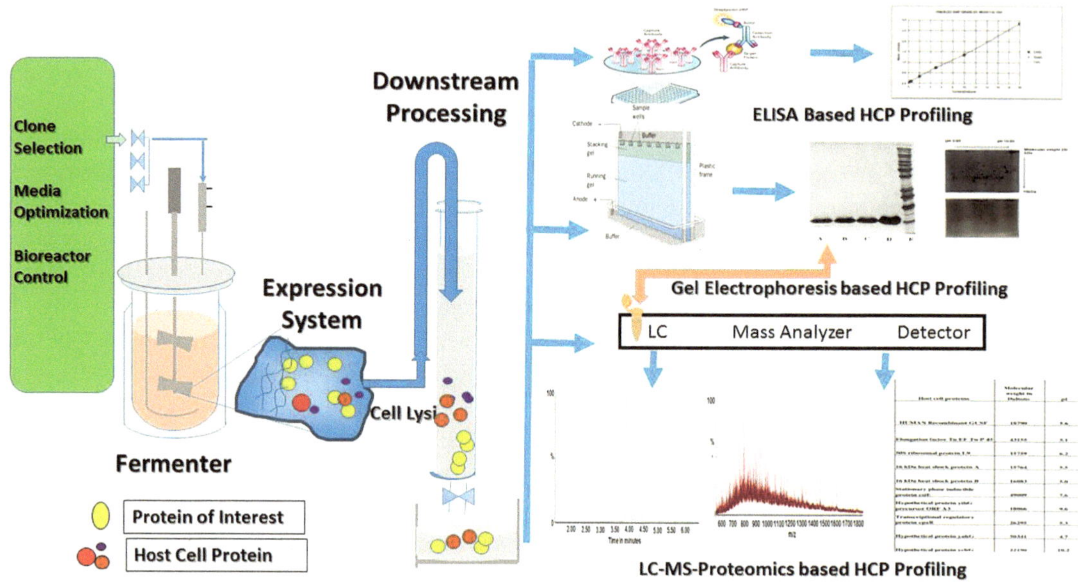

Fig. 7.3 Analytical techniques used for HCP profiling in biotherapeutics. The complexity of the processes that are used to manufacture protein therapeutics along with the raw materials that they use results in co-purification of some HCPs that bear close resemblance to the protein of interest. Broadly, three platforms are used for monitoring HCP. ELISA-based methods help in quantifying the amount of HCP present but do not provide details about the identity of unwanted proteins. Gel-based assays have higher coverage than ELISA but lower dynamic range. Mass spectrometry-based proteomic platforms enable detailed characterization of biologics, both qualitatively and quantitatively, with a dynamic range much higher than the other two platforms

7.4 Functional Characterization

While it is a necessity that a biosimilar is highly similar to the innovator product in terms of its structural makeup, equivalent bioactivity is also a necessary prerequisite and needs to be verified with appropriate functional assays. Label-free detection technologies like surface plasma resonance spectroscopy (Hulse and Cox 2013a) and biolayer interferometry-based (Abdiche et al. 2008) ligand-binding assays are becoming popular tools for biosimilar development for detection, monitoring, and characterization of bimolecular interactions. Apart from this, these assays also allow estimation of potency of biosimilar preparations. Unlike the traditional methods like ELISA and flow cytometry that involve labeling of the molecule of interest with fluorescent dyes, radioisotopes, and/or epitope tags, the label-free technologies are based on measurement of a molecule's intrinsic property, thereby allowing direct, real-time measurement of biomolecules in a high-throughput manner (Ray et al. 2010; Hulse and Cox 2013b). This helps in overcoming of the issue of alteration of surface characteristics and confirmation of the molecule exhibiting interference with the binding site with the traditional binding characterization assays. Further, BLI-/SPR-based ligand-binding assays are suitable for testing of the drug before administration to preclinical or clinical recipients, as well as for testing of an active drug in preclinical or clinical recipient's sera. The assays would involve assessment of the activity of the drug (in terms of its disposition in the body via pharmacokinetic and pharmacodynamic measurements) in the sera as well as activity of potentially induced host cell neutralizing antidrug antibodies (immunogenicity assessment) (Challener 2015). This is of very high significance as there are only few cell-based bioassays that show good agreement to the results obtained from ELISA/SPR/BLI based analysis for sera that is derived from preclinical or clinical recipients (Kuwabara et al. 1996).

However, there is a unanimous opinion that wherever cell-based assays exists for making such assessments (e.g., erythropoietin, Granulocyte

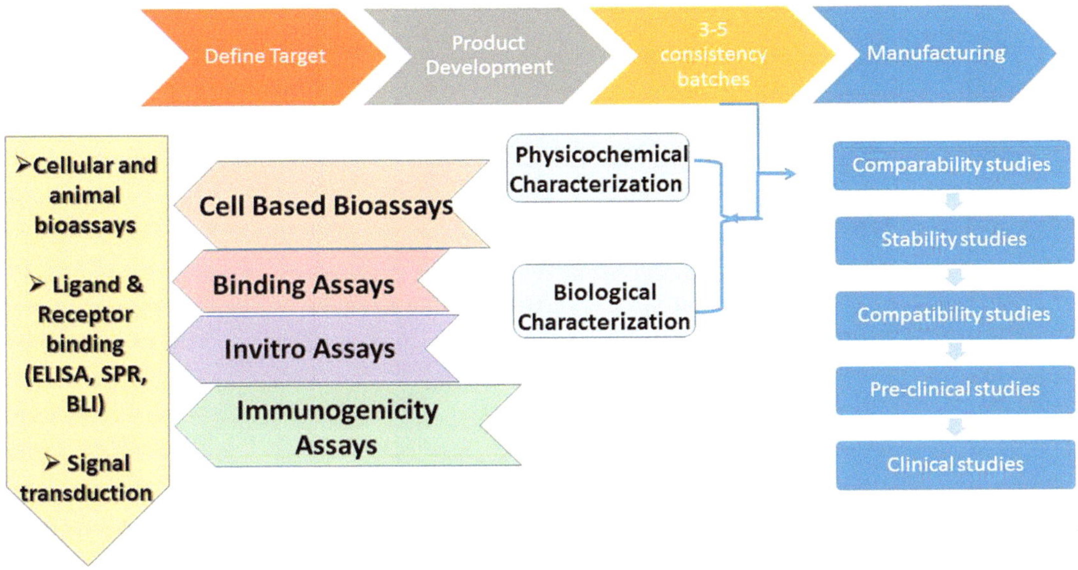

Fig. 7.4 Biosimilar development chain highlighting functional characterization. Functional characterization is routinely undertaken from early drug discovery for measurement of consistency between batches of drug produced and to establish equivalence between the biosimilar and the corresponding innovator product. While cell-based assays are the most common used platform for potency assessment, in vitro binding assays using label-free techniques like surface plasma resonance and bio-layer interferometry are gaining popularity as they allow detection, monitoring, and characterization of bimolecular interactions. The ability to couple SPR with mass spectrometry gives possibility of proteomic-based functional characterization studies for biotherapeutics

colony stimulating factor), it should be resorted as the first choice of biosimilarity testing as the former is more biomimetic to the ligand's mechanism of action as compared to ligand-binding assays and tends to be more sensitive to small differences or changes (Challener 2015). Figure 7.4 provides an overview of the biosimilar development chain and stages therein where functional characterization studies are to be carried out.

Finally, Table 7.2 presents a summary of recent experimental findings of biotherapeutic characterization using proteomics.

7.5 Future Perspective

The expansion of biosimilar market worldwide will primarily depend on addition of more sophisticated platforms to our analytical armory that can be exploited for characterization of biologics. Proteomics is an indispensable tool in the characterization of these biosimilars which would regulate the efficacy of these therapeutic products. Based on the developments outlined in this chapter, we can say that mass spectrometry-based proteomic approaches are and will continue to be a significant enabler in this regard. While existing tools have been successful in qualitative and quantitative measurement of protein structure and function, we are far off from obtaining a complete fingerprint of a biotherapeutic. The latter is likely to revolutionize our ability in assessing the impact of the various quality attributes of a biotherapeutic on its clinical safety and efficacy. Equally importantly, this will continue to reduce our reliance on clinical studies for making this assessment and thereby lower the cost of healthcare. This is necessary if we wish to make these biotherapeutics accessible to the various developing and underdeveloped economies of the world, especially to provide them with more affordable options for therapy. We think that academia, government agencies, and the biotech industry need to continue to work together to achieve this vision.

Table 7.2 Summary of literature on characterization of biotherapeutics using proteomic tools

Category	Aim	Tools used	Comments	Reference
Intact level	Primary sequence and molecular identity assessment of adalimumab biosimilar	LC/MS/MS MALDI-TOF-MS	Biosimilarity established with array of analytical tools	Bandyopadhyay et al. (2015)
Intact level	Structural characterization	RP-LC-FTMS With ETD function	Proteolysis achieved in 30 min, ~70 % sequence coverage, identification of oxidation sites	Fornelli et al. (2014)
Intact level	Structural analysis of mAbs	Extended bottom-up proteomics (using secreted aspartic protease 9/sap9)	~100 % and 99 % sequence coverage for light chain and heavy chain, respectively, in single LC run, low artifacts, antibody cocktail analysis	Laskay et al. (2014)
Higher order	Determination of disulfide shuffling in biopharmaceuticals	MALDI-TOF	Automated profiling and identification of disulfide bond shuffling in time-effective manner	Wiesner et al. (2015)
Higher order	Determination of PTMs and degradation hot spots during biosimilar mAb manufacture	CE-ESI-MS	Comprehensive assessment of PTMs and degradation hot spots	Lew et al. (2015)
Higher order	Characterization of higher-order structure and structural dynamics of therapeutic proteins	HDX-MS with subzero temperature chromatography	Protein characterization under physiological conditions with single-residue resolution; no disulfide bond shuffling during analysis	Pan et al. (2014)
Glycosylation	N- and O-linked glycosylation analysis of a fusion therapeutic protein (etanercept)	LC-qTOF with ETD function	N- and O-linked glycan profiling in terms of sequence and linkage information	Houel et al. (2014)
Glycosylation	Characterization of posttranslational modifications	LC-TOF-MS with ETD function	Superior performance of ETD than CID in terms of sequence coverage	Williams et al. (2013)
HCP profiling	HCP profiling in drug product/substance	1-D LC/MS	HCP enrichment by product depletion followed by multidimensional LC/MS enhanced the detection and identification of HCPs	Thompson et al. (2014)
		2-D LC/MS	Potential for basing the method for targeted immunological assay development	

References

Abdiche Y, Malashock D, Pinkerton A, Pons J (2008) Determining kinetics and affinities of protein interactions using a parallel real-time label-free biosensor, the Octet. Anal Biochem 377:209–217

Ayoub D, Jabs W, Resemann A, Evers W, Evans C, Main L et al (2013) Correct primary structure assessment and extensive glyco-profiling of cetuximab by a combination of intact, middle-up, middle-down and bottom-up ESI and MALDI mass spectrometry techniques. MAbs 5:699–710

Bandyopadhyay S, Mahajan M, Mehta T, Singh AK, Parikh A, Gupta AK et al (2015) Physicochemical and functional characterization of a biosimilar adalimumab ZRC-3197. Biosimilars 5:1–18

Beck A, Sanglier-Cianférani S, Van Dorsselaer A (2012) Biosimilar, biobetter, and next generation antibody characterization by mass spectrometry. Anal Chem 84:4637–4646

Beck A, Diemer H, Ayoub D, Debaene F, Wagner-Rousset E, Carapito C et al (2013) Analytical characterization of biosimilar antibodies and Fc-fusion proteins. TrAC – Trends Anal Chem 48:81–95

Beck A, Debaene F, Diemer H, Wagner-Rousset E, Colas O, Van Dorsselaer A et al (2015) Cutting-edge mass spectrometry characterization of originator, biosimilar and biobetter antibodies. J Mass Spectrom 50:285–297

Berkowitz SA, Engen JR, Mazzeo JR, Jones GB (2012) Analytical tools for characterizing biopharmaceuticals and the implications for biosimilars. Nat Rev Drug Discov 11:527–540

Burton DR, Dwek RA (2006) Immunology. Sugar determines antibody activity. Science 313:627–628

Butler M (2005) Animal cell cultures: recent achievements and perspectives in the production of biopharmaceuticals. Appl Microbiol Biotechnol 68:283–291

Challener CA (2015) Ligand-binding assays and the determination of biosimilarity. Biopharm Int 28(1):36–39

Champion K, Madden H, Dougherty J, Shacter E (2005) Defining your product profile and maintaining control over it, part 2. Bioprocess Int 3:52–57

Chen SL, Wu SL, Huang LJ, Huang JB, Chen SH (2013) A global comparability approach for biosimilar monoclonal antibodies using LC-tandem MS based proteomics. J Pharm Biomed Anal 80:126–135

Chirino AJ, Mire-Sluis A (2004) Characterizing biological products and assessing comparability following manufacturing changes. Nat Biotechnol 22:1383–1391

Cottrell JS (2011) Protein identification using MS/MS data. J Proteomics 74:1842–1851

Damen CWN, Chen W, Chakraborty AB, van Oosterhout M, Mazzeo JR, Gebler JC et al (2009) Electrospray ionization quadrupole ion-mobility time-of-flight mass spectrometry as a tool to distinguish the lot-to-lot heterogeneity in N-glycosylation profile of the therapeutic monoclonal antibody trastuzumab. J Am Soc Mass Spectrom 20:2021–2033

Debaene F, Wagner-Rousset E, Colas O, Ayoub D, Corvaïa N, Van Dorsselaer A et al (2013) Time resolved native ion-mobility mass spectrometry to monitor dynamics of igg4 fab arm exchange and 'bispecific' monoclonal antibody formation. Anal Chem 85:9785–9792

Doneanu CE, Xenopoulos A, Fadgen K, Murphy J, Skilton SJ, Prentice H et al (2012) Analysis of host-cell proteins in biotherapeutic proteins by comprehensive online two-dimensional liquid chromatography/mass spectrometry. MAbs 4:24–44

Eon-Duval A, Broly H, Gleixner R (2012) Quality attributes of recombinant therapeutic proteins: an assessment of impact on safety and efficacy as part of a quality by design development approach. Biotechnol Prog 28:608–622

Fornelli L, Ayoub D, Aizikov K, Beck A, Tsybin YO (2014) Middle-down analysis of monoclonal antibodies with electron transfer dissociation orbitrap fourier transform mass spectrometry. Anal Chem 86:3005–3012

Goochee CF, Gramer MJ, Andersen DC, Bahr JB, Rasmussen JR (1991) The oligosaccharides of glycoproteins: bioprocess factors affecting oligosaccharide structure and their effect on glycoprotein properties. Biotechnology (N Y) 9:1347–1355

Harvey SR, MacPhee CE, Barran PE (2011) Ion mobility mass spectrometry for peptide analysis. Methods 54:454–461

Haselberg R, Brinks V, Hawe A, De Jong GJ, Somsen GW (2011) Capillary electrophoresis-mass spectrometry using noncovalently coated capillaries for the analysis of biopharmaceuticals. Anal Bioanal Chem 400:295–303

Hillenkamp F, Beavis RC, Brian T (1991) Matrix-assisted laser desorption/ionization mass spectrometry of biopolymers. Anal Chem 63:1193A–1203A

Hirsch BR, Lyman GH (2014) Biosimilars: a cure to the U.S. health care cost conundrum? Blood Rev 28:263–268

Hogwood CE, Bracewell DG, Smales CM (2014) Measurement and control of host cell proteins (HCPs) in CHO cell bioprocesses. Curr Opin Biotechnol 30C:153–160

Houde D, Peng Y, Berkowitz SA, Engen JR (2010) Post-translational modifications differentially affect IgG1 conformation and receptor binding. Mol Cell Proteomics 9:1716–1728

Houde D, Berkowitz SA, Engen JR (2011) The utility of hydrogen/deuterium exchange mass spectrometry in biopharmaceutical comparability studies. J Pharm Sci 100:2071–2086

Houel S, Hilliard M, Yu YQ, McLoughlin N, Martin SM, Rudd PM et al (2014) N-and O-glycosylation analysis of etanercept using liquid chromatography and quadrupole time-of-flight mass spectrometry equipped with electron-transfer dissociation functionality. Anal Chem 86:576–584

Hulse J, Cox C (2013a) In vitro functional testing methods for monoclonal antibody biosimilars. Bioprocess Int 11:24–28

Hulse J, Cox C (2013b) SPR & flow cytometry for biosimilars. Available at http://www.contractpharma.com/issues/2013-05/view_features/spr-flow-cytometry-for-biosimilars. Accessed on 30 May 2016

Jin M, Szapiel N, Zhang J, Hickey J, Ghose S (2010) Profiling of host cell proteins by two-dimensional difference gel electrophoresis (2D-DIGE): implications for downstream process development. Biotechnol Bioeng 105:306–316

Kálmán-Szekeres Z, Olajos M, Ganzler K (2012) Analytical aspects of biosimilarity issues of protein drugs. J Pharm Biomed Anal 69:185–195

Kaltashov IA, Bobst CE, Abzalimov RR (2013) Mass spectrometry-based methods to study protein architecture and dynamics. Protein Sci 22:530–544

Konermann L, Pan J, Liu Y-H (2011) Hydrogen exchange mass spectrometry for studying protein structure and dynamics. Chem Soc Rev 40:1224–1234

Konermann L, Vahidi S, Sowole MA (2014) Mass spectrometry methods for studying structure and dynamics of biological macromolecules. Anal Chem 86:213–232

Kunkel JP, Jan DCH, Butler M, Jamieson JC (2000) Comparisons of the glycosylation of a monoclonal antibody produced under nominally identical cell culture conditions in two different bioreactors. Biotechnol Prog 16:462–470

Laskay UA, Srzentić K, Monod M, Tsybin YO (2014) Extended bottom-up proteomics with secreted aspartic protease Sap9. J Proteomics 10:20–31

Levy NE, Valente KN, Choe LH, Lee KH, Lenhoff AM (2014) Identification and characterization of host cell protein product-associated impurities in monoclonal antibody bioprocessing. Biotechnol Bioeng 111:904–912

Lew C, Gallegos-Perez J-L, Fonslow B, Lies M, Guttman A (2015) Rapid level-3 characterization of therapeutic antibodies by capillary electrophoresis electrospray ionization mass spectrometry. J Chromatogr Sci 53:443–449

Lodowski DT, Palczewski K, Miyagi M (2010) Conformational changes in the G protein-coupled receptor rhodopsin revealed by histidine hydrogen-deuterium exchange. Biochemistry 49:9425–9427

Lopez MF, Berggren K, Chernokalskaya E, Lazarev A, Robinson M, Patton WF (2000) A comparison of silver stain and SYPRO ruby protein gel stain with respect to protein detection in two-dimensional gels and identification by peptide mass profiling. Electrophoresis 21:3673–3683

Lorenzen K, van Duijn E (2010) Native mass spectrometry as a tool in structural biology. Curr Protoc Protein Sci 1–17

Lynaugh H, Li H, Gong B (2013) Rapid Fc glycosylation analysis of Fc fusions with IdeS and liquid chromatography mass spectrometry. MAbs 5:641–645

Morelle W, Michalski J-C (2007) Analysis of protein glycosylation by mass spectrometry. Nat Protoc 2:1585–1602

Myszka DG, Rich RL (2000) Implementing surface plasmon resonance biosensors in drug discovery. Pharm Sci Technol Today 3:310–317

Pan J, Borchers CH (2014) Top-down mass spectrometry and hydrogen/deuterium exchange for comprehensive structural characterization of interferons: implications for biosimilars. Proteomics 14:1249–1258

Pan J, Zhang S, Parker CE, Borchers CH (2014) Subzero temperature chromatography and top-down mass spectrometry for protein higher-order structure characterization: method validation and application to therapeutic antibodies. J Am Chem Soc 136:13065–13071

Pritchard C, O'Connor G, Ashcroft AE (2013) The role of ion mobility spectrometry-mass spectrometry in the analysis of protein reference standards. Anal Chem 85:7205–7212

Rathore AS (2009a) Follow-on protein products: scientific issues, developments and challenges. Trends Biotechnol 27:698–705

Rathore AS (2009b) Roadmap for implementation of quality by design (QbD) for biotechnology products. Trends Biotechnol 27:546–553

Rathore AS, Bhambure R (2014) Establishing analytical comparability for 'biosimilars': filgrastim as a case study. Anal Bioanal Chem 406:6569–6576

Ray S, Mehta G, Srivastava S (2010) Label-free detection techniques for protein microarrays: prospects, merits and challenges. Proteomics 10:731–748

Rosati S, Yang Y, Barendregt A, Heck AJR (2014) Detailed mass analysis of structural heterogeneity in monoclonal antibodies using native mass spectrometry. Nat Protoc 9:967–976

Schenauer MR, Flynn GC, Goetze AM (2012) Identification and quantification of host cell protein impurities in biotherapeutics using mass spectrometry. Anal Biochem 428:150–157

Schenauer MR, Flynn GC, Goetze AM (2013) Profiling the effects of process changes on residual host cell proteins in biotherapeutics by mass spectrometry. Biotechnol Prog 29:951–957

Singleton CA (2014) MS in the analysis of biosimilars. Bioanalysis 6:1627–1637

Srgel F, Lerch H, Lauber T (2010) Physicochemical and biologic comparability of a biosimilar granulocyte colony-stimulating factor with its reference product. BioDrugs 24:347–357

Tait AS, Hogwood CEM, Smales CM, Bracewell DG (2012) Host cell protein dynamics in the supernatant of a mAb producing CHO cell line. Biotechnol Bioeng 109:971–982

Tang L, Sundaram S, Zhang J, Carlson P, Matathia A, Parekh B et al (2013) Conformational characterization of the charge variants of a human IgG1 monoclonal antibody using H/D exchange mass spectrometry. MAbs 5:114–125

Thompson JH, Chung WK, Zhu M, Tie L, Lu Y, Aboulaich N et al (2014) Improved detection of host cell proteins (HCPs) in a mammalian cell-derived antibody drug using liquid chromatography/mass spectrometry in conjunction with an HCP-enrichment strategy. Rapid Commun Mass Spectrom 28:855–860

Vahidi S, Stocks BB, Konermann L (2013) Structure in the gas phase. Anal Chem 85:10471–10478

Velez-Suberbie ML, Tarrant RDR, Tait AS, Spencer DIR, Bracewell DG (2013) Impact of aeration strategy on CHO cell performance during antibody production. Biotechnol Prog 29:116–126

Visser J, Feuerstein I, Stangler T, Schmiederer T, Fritsch C, Schiestl M (2013) Physicochemical and functional comparability between the proposed biosimilar rituximab GP2013 and originator rituximab. BioDrugs 27:495–507

Walsh G (2014) Biopharmaceutical benchmarks 2014. Nat Biotechnol 32:992–1000

Wang J, Chow SC (2012) On the regulatory approval pathway of biosimilar products. Pharmaceuticals 5:353–368

Warnke S, Von Helden G, Pagel K (2013) Protein structure in the gas phase: the influence of side-chain microsolvation. J Am Chem Soc 135:1177–1180

Wei H, Mo J, Tao L, Russell RJ, Tymiak AA, Chen G et al (2014) Hydrogen/deuterium exchange mass spectrometry for probing higher order structure of protein therapeutics: methodology and applications. Drug Discov Today 19:95–102

Wiesner J, Resemann A, Evans C, Suckau D, Jabs W (2015) Advanced mass spectrometry workflows for analyzing disulfide bonds in biologics. Expert Rev Proteomics 12:115–123

Williams JP, Pringle S, Richardson K, Gethings L, Vissers JPC, De Cecco M et al (2013) Characterisation of glycoproteins using a quadrupole time-of-flight mass spectrometer configured for electron transfer dissociation. Rapid Commun Mass Spectrom 27:2383–2390

Wilm M (2011) Principles of electrospray ionization. Mol Cell Proteomics 10:M111–M009407

Wu SL, Hühmera FR, Hao Z, Karger BL (2007) On-line LC-MS approach combining collision-induced dissociation (CID), electron-transfer dissociation (ETD), and CID of an isolated charge-reduced species for the trace-level characterization of proteins with post-translational modifications. J Proteome Res 6:4230–4244

www.alliedmarketresearch.com/biosimilars-market

Xie H, Chakraborty A, Ahn J, Yu YQ, Dakshinamoorthy DP, Gilar M et al (2010) Rapid comparison of a candidate biosimilar to an innovator monoclonal antibody with advanced liquid chromatography and mass spectrometry technologies. MAbs 2:379–394

Zhang A, Qi W, Singh SK, Fernandez EJ (2011) A new approach to explore the impact of freeze-thaw cycling on protein structure: hydrogen/deuterium exchange mass spectrometry (HX-MS). Pharm Res 28:1179–1193

Zhong Y, Hyung S-J, Ruotolo BT (2012) Ion mobility–mass spectrometry for structural proteomics. Expert Rev Proteomics 9:47–58

Zhu G, Sun L, Thompson JH, Kuntumalla S, Lin HY, Larkin CJ, McGivney IV JB, Dovichi NJ (2016) Capillary zone electrophoresis tandem mass spectrometry detects low concentration host cell impurities in monoclonal antibodies. Electrophoresis 37:616–622

Regulatory Norms and Intellectual Property Rights for Biomarker Research

8

Tumpa Raj Das, Apoorva Venkatesh, Apurva Atak, Shabarni Gupta, and Prasad B. Phapale

Abstract

The healthcare industry has hugely benefited from the advent of biomarkers in diagnostics. The process of translation of a biomarker from "bench to bedside," however, involves various key steps which are monitored by autonomous bodies to ensure that the biomarker/biosimilars under consideration are safe and meet the regulatory guidelines established by them. This chapter aims to provide an overview of global regulatory bodies and discusses the various norms required for commercialization of biomarkers, with special focus on the role of proteomic-based tools to help study the efficacy of these biomarkers. Another facet, concurrent with the post discovery-based endeavors from regulatory bodies, is the protection of intellectual property rights of a researcher's discovery-based work. Patent claims have repeatedly been under the scrutiny of legislative bodies due to controversies on subjects like patent eligibility resulting in the impediment of scientific progress and research. This has consequently resulted in restricting the disputed product's use in diagnoses and other applications. Additionally, this chapter overviews the nuances of patent filing to protect the intellectual property rights of researchers involved in the discovery of biomarkers prior to their commercialization.

Keywords

Biomarker ICH guidelines • ICH E16 • FDA review • Biomarker assays • Biomarker evaluation • US patent law • Biomarker patents • Composition of matter claims • Method claims • Infectious and noninfectious diseases

T.R. Das • A. Venkatesh • A. Atak • S. Gupta
Department of Biosciences and Bioengineering, Indian Institute of Technology Bombay, Powai, Mumbai 400076, Maharashtra, India

P.B. Phapale (✉)
Structural and Computational Biology Unit, European Molecular Biology Laboratory (EMBL), Meyerhofstraße 1, 69117 Heidelberg, Germany
e-mail: prasad.phapale@embl.de

S. Srivastava (ed.), *Biomarker Discovery in the Developing World: Dissecting the Pipeline for Meeting the Challenges*, DOI 10.1007/978-81-322-2837-0_8

8.1 Regulatory Bodies in Pharmaceutical Sector

The beginning of the twenty-first century has been marked by the surge of research and development in the area of biopharmaceuticals (Witcher Odum 2012). While food, shelter, and clothing were considered to be the basic needs for human survival, in today's time, this definition of "basic needs" has been broadly extended to sustainable healthcare and therapeutics. Considering the presence and impact of these biopharmaceuticals in the life of common man, strict regulation of these healthcare products from the research phase to the commercial phase is an imperative. This need for safety in drug administration has given rise to regulatory bodies. The role of these bodies was to empower commercial drug manufacturers with a set of guidelines which could be followed for ensuring the safety and efficacy of the drug being commercialized.

The case of the Thalidomide tragedy in the 1960s is a popular example emphasizing the prevalence of drug safety and regulation mishaps occurring in that era which eventually led to the emergence of ICH and other regulatory bodies during the 1950s and 1960s. Thalidomide was prescribed to pregnant females as a treatment toward curing morning sickness. However, in the 1960s, a major side effect of the drug resulted in a rare disorder called Phocomelia. Thalidomide-induced Phocomelia caused infants to be born with congenital disorders which manifested as aural, facial, and limb abnormalities. Dr. Frances Oldham Kelsey, an employee of the Food and Drug Administration (FDA), exposed the drug's side effects only after extensive safety studies which were performed by the manufacturing company for its clearance in the US markets. This drug was removed from the market in 1962; however, around 10,000 children had fallen prey to the side effects of this drug, and about 40 % of the affected kids died within the first year of their birth (Thalidomide crisis and drug regulation, Emory library [Online]).

Proper regulatory clearance before marketing any drug had, thus, been initiated following several such incidences. In case of biopharmaceutical drug delivery (with or without biomarkers), the significance of regulatory clearance is even more, in consideration of the unprecedented liability and accountability that drug manufacturers would have in terms of public health and safety. Thus, the role of guidelines laid down by regulatory bodies is important to ensure safety of biomarkers and biologics to the end consumer.

8.1.1 Overview of Various Regulatory Bodies Worldwide

In the process of drug research and development, and its journey from discovery to market, quality, safety, and efficacy remain the prime focus of all drug manufacturers and regulatory bodies. Drug manufacturers follow certain specifications to bring up a final product in order to come up with a stand-alone or a combination drug which is based on quality and therapeutic efficiency and safe for patients of various age groups, race, sex, etc. depending upon the prescribed dosage type and regimen.

These specifications, which guide the manufacturing companies through all the stages of formulation of drugs and drug products, are collectively referred to as guidelines, and the governing bodies which regulate these guidelines are known as regulatory bodies. Thus, abiding by the stringent guidelines laid by the regulatory bodies is more than a mere obligation to drug manufacturers.

In today's time, in order to launch a drug in any given country, it is mandatory to get the regulatory clearance from the concerned regulatory body depending upon the market of the drug. Many national and international regulatory bodies have come into existence in the last few decades. However, the most widely noted and followed regulatory bodies worldwide are ICH and WHO (international), US-FDA (United States), EMA (European), DCGI (CDSCO India), PMDA (Japan), and SFDA (China) (Pharmaweb, government, and regulatory bodies) depending on the nature of the product being manufactured.

The International Conference on Harmonization of Technical Requirements for

Registration of Pharmaceuticals for Human Use (ICH) originated in 1990 and is an international body of regulatory authorities which has evolved gradually. The ICH guidelines have played an important role in drug research and development. They have been followed worldwide in pharmaceutical and biopharmaceutical sectors right from the stages of developing a drug to its commercialization.

The ICH guidelines are categorized broadly into three classes, dealing with efficacy, safety, and quality. The nature and number of ICH guidelines are given in Table 8.1. Multidisciplinary category contains guidelines for those products which do not fall in quality, safety, and efficacy bars. The drug manufacturers follow the ICH guidelines for all the steps beginning with product initiation, preformulation, formulation, analytical studies, stability, pilot scale, validation, etc. Apart from the general guidelines of ICH, most countries, in fact all markets, have their own specifications which are expected to be followed by the drug manufacturers for marketing their products in a particular country (ICH guidelines [Online]).

US Food and Drug Administration (US-FDA) is the regulatory body for food, drug, and public health in the United States. The US-FDA also supports various programs for the innovation of new drugs which are effective, safer, and more economical. The FDA undertakes many such responsibilities to ensure improvement of overall public health. The FDA frames various guidances for drug safety, and it also has a list of FDA-approved drugs mentioned both in the "drugs@FDA" and "the Orange Book" data for reference by various generic manufacturers. It also displays information about a drug's status on its approval as a reference listed drug (RLD), which is highly beneficial to companies dealing with drug development especially for generic formulations (US-FDA [Online]).

European Medicines Agency (EMA or EMEA) is the regulatory body responsible for health welfare in the European Union and is headquartered in London. It regulates drug safety and efficacy of all drugs that are to be marketed in the European Union. In compliance with the ICH guidelines, the EMA guidelines play significant role as a large number of drug manufacturers target the European countries to export of their drugs. The EMA guidelines cover various aspects of formulation and development, viz., quality of active pharmaceutical ingredient (API), specifications for impurities in API, or product during development, stability, packaging, post approval drug changes, etc. (Pharmaceutical development, EMA [Online]).

Drug Controller General of India (DCGI) is under the control of CDSCO of India (The Central Drugs Standard Control Organization), which is the Central Drug Authority for carrying out functions assigned to the Central Government under the Drugs and Cosmetics Act. The DCGI is responsible to sanction or reject the marketing of various drugs, with respect to clinical studies of new drugs, bioavailability and bioequivalence studies, quality and safety, etc. As regards to biologics, many guidelines for biosimilars, vaccine manufacturing, stem cell and stem cell-based products, etc. have been generated by the CDSCO (Biologicals, CDSCO [Online]).

Pharmaceuticals and Medical Devices Agency (PMDA), Japan, is a Japanese regulatory agency, which closely works with their Ministry of Health, Labor, and Welfare. The PMDA, like other regulatory bodies, controls complete public health in Japan ensuring quality, safety, and efficacy. The drug approval in Japan is generally based on the clinical studies done in Japan. The timeline for drug approval, i.e., clinical trial's approval in Japan, is much longer than in other countries, and therefore many laws are reformed to avoid competitive market drawbacks due to such delays (Satoru Nagasaki et al. 2008; PMDA [Online]).

8.1.2 ICH Guidelines for Biomarkers

The FDA pharmacogenomic guidance defines a valid biomarker as "a biomarker that is measured in an analytical test system with well established performance characteristics and for which there is an established scientific framework or body of evidence that elucidates the physiologic,

Table 8.1 An overview of ICH guidelines

Category	Guideline number	Topics covered
Quality	Q1–Q12	Stability, analytical validation, impurities, pharmacopeias, quality of biotechnological products, specifications, good manufacturing practice, pharmaceutical development, quality risk management, pharmaceutical quality system, development and manufacture of drug substances, and life cycle management, respectively
Safety	S1–S11	Carcinogenicity studies, genotoxicity studies, toxicokinetics and pharmacokinetics, toxicity testing, reproductive toxicology, biotechnological products, pharmacology studies, immunotoxicology studies, nonclinical evaluation for anticancer pharmaceuticals, photosafety evaluation, and nonclinical safety testing, respectively
Efficacy	E1–E18	Clinical safety for drugs used in long-term treatment, pharmacovigilance, clinical study reports, dose-response studies, ethnic factors, good clinical practice, clinical trials in geriatric population, general considerations for clinical trials, statistical principles for clinical trials, choice of control group in clinical trials, clinical trials in pediatric population, clinical evaluation by therapeutic category, clinical evaluation, definitions in pharmacogenetics/pharmacogenomics, qualification of genomic, multiregional clinical trials, and genomic methodologies, respectively
Multidisciplinary	M1–M8	MedDRA terminology, electronic standards, nonclinical safety studies, common technical document, data elements and standards for drug dictionaries, gene therapy and genotoxic impurities, Electronic Common Technical Document (eCTD)

ICH guidelines for quality, safety, efficacy, and multidisciplinary http://www.ich.org/products/guidelines.html

toxicologic, pharmacologic, or clinical significance of the test results." The biomarkers as already described in the introductory chapters of this book are highly beneficial in early diagnosis of diseases and also provide better treatment and ease monitoring disease progression (Steven Kozlowski; FDA perspectives on Biomarkers).

The E15 and E16 are the ICH guidelines designated for regulatory requirement for biomarkers. As the name says, the E15 guideline describes the various aspects of genomic biomarkers, and E16 guideline deals with context, structure, and other aspects of biomarkers for its development as a drug or its presence in a biotechnological product. The E16 guideline was initially assigned to genomic biomarkers, but it also mentions that the principles remain the same for other biomarkers, viz., proteomics, imaging, and other non-genomic biomarkers. The scope of the guidance is preparation of data for submission of R&D evaluation and various clinical and post clinical assessments.

The data for biomarker qualification submission needs to be in compliance with the common technical document (CTD) format and is accordingly required for the submission and review. New drug application (NDA), biologics license application (BLA), and a marketing authorization application (MAA) are significant documents required for biomarker regulatory approval processes based on the type of submission and also depending upon the request by regulatory authorities. Biomarkers are qualified for a specific context of use. The context of qualification depends on various criteria like stand-alone use of biomarkers or in combination with a biotechnological product or in process. For a biomarker qualification submission, specific sections should be included in the structure for subsequent approval as mentioned in Table 8.2 (FDA E15 2008 and E16 2011 guidelines [Online]).

Table 8.2 Brief description of sections required to be included for submission of a biomarker qualification

Section	Contents	Brief description
Section 1	Regional administrative information	Specific to region mainly contains application forms and cover letters for the biomarker submission
Section 2	Summaries	Overall biomarker qualification, i.e.,biomarker description, use, diseases, experimental setting, experimental design, and context of use based on pharmacology, toxicology, safety, and disease
Section 3	Quality reports	Quality and manufacturing data of the biomarker drug or the biotechnology product not expected independent from NDA, BLA, or MAA
Section 4	Non-clinical reports	Analytical assay development and validation reports and nonclinical in vitro and in vivo reports
Section 5	Clinical reports	Analytical assay development and validation reports and clinical pharmacology, efficacy, and/or safety reports

E16 guidelines of biomarkers related to drug or biotechnology product development: content, structure, and format of qualification submissions Guidance for Industry (2011) ICH, FDA, Hampshire

8.2 Regulatory Perspective on Translating Biomarkers from Bench to Bedside

Remarkable research efforts have been made over the years to discover new biomarkers for disease diagnosis and therapeutic monitoring. However, the failure rate of such efforts has resulted in a downfall of these efforts where the number of biomarkers eventually reaching clinical practice is very low. Füzéry et al. summarized those few FDA-approved tumor protein markers, which are presently used in clinical practice (Füzéry et al. 2013).

Among several challenges, the most potential cause of this failure has been highlighted by Vidal et al. as the lack of knowledge about regulatory requirements among the scientists working in biomarker research (Vidal et al. 2012). We have traced the importance of awareness about regulatory norms among scientists working in biomarkers and multi-omics research areas which have been described below. This will further facilitate the direction of research in translating biomarkers from bench to bedside. We will discuss regulatory framework and norms particularly for proteomic biomarker discovery and its scope in other omics research areas.

8.2.1 Translating Proteomic Biomarkers to Clinical Diagnostics

The translation of proteomic biomarkers from discovery to clinical diagnostics has several challenges and issues. Such challenges are well addressed by academic scientists, government authorities and regulatory bodies, as well as companies involved in manufacturing diagnostic tools, assay kits, and instruments. Recently the mass spectrometry (MS)-based platforms are becoming one of the most promising tools for biomarker analysis. Figure 8.1 describes the overview of biomarker development cycle from discovery research to regulatory approval through validation stages. The patenting and intellectual property protection can be (not necessarily) part of the process depending on the type of assay product or process. Biomarker patenting will be discussed more in the next section of this chapter.

For proteomics, specific regulatory norms have to be considered due to their experimental complexities, uniqueness in analyte measurements, and instrumentation. Hence the FDA regulatory framework for in vitro diagnostic tests to regulate proteomics and other "omics" tests needs some special considerations in the light of new developments (Lee et al. 2011; Li et al. 2011; Lopez et al. 2012; Shreeve et al. 2012; Haleem and Timothy 2013).

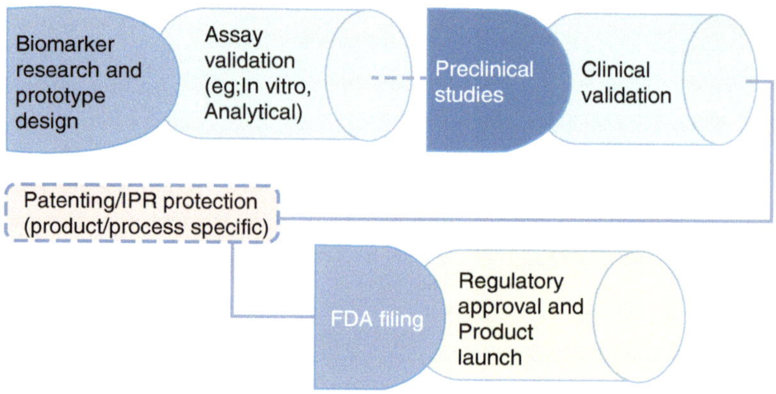

Fig. 8.1 Overview of biomarker development cycle from discovery to regulatory approval product launch

8.2.2 Steps in FDA Review Process

The FDA review for a premarket approval application (PMA) is a four-step review process, which includes (based on Draft Guidance for Industry and FDA Staff Medical Devices and Lee et al. 2011:

1. Filing review for fullness of submitted application in terms of administrative and limited scientific perspective. Table 8.2 summarizes some general sections required to be included for submission of a biomarker qualification.
2. Quality system review is meticulous and scientific which needs regulatory evaluation by relevant FDA experts.
3. Panel review requires recommendation by the designated advisory committee.
4. Final decision by FDA based on documentation and notification to applicants.

8.2.3 Classification for Proteomic Biomarker Assays

Since FDA considers protein-based biomarker assays as medical devices that are used for diagnosis, prognosis, screening, therapeutic monitoring, and medical treatment, proteomic biomarkers follow the same regulatory framework and process.

The office of In Vitro Diagnostics and Radiological Health (OIR) regulates such assays, which are part of laboratory diagnostic tests and in vitro diagnostic devices (IVDs). Center for Biologics Evaluation and Research (CBER) is a department within FDA for human biological products which regulates some diagnostic medical devices dealing with HIV, human leukocyte antigen, and blood analysis.

FDA divides medical devices in three classes depending on safety, risk of errors on patient's health, and their intended use (Lee et al. 2011; Li et al. 2011; Lopez et al. 2012; Haleem and Timothy 2013):

Class I devices are mostly considered as low risk and generally such devices are exempted from premarket review process.

Class II devices are supposed to carry moderate risk and usually require a premarket notification to demonstrate their equivalence in terms of safety and effectiveness with legally marketed devices (21CFR 807.92(a)(3)) which are exempted from premarket approval (PMA).

Class III devices are high risk, which need considerable in-depth review of the safety and effectiveness with experimental and well-documented evidences. The test errors in such devices or assays may cause major health risks to the patients.

8.2.3.1 Investigational Device Exemption (IDE)

The Federal Food, Drug, and Cosmetic Act (FD&C Act) Section 520 (g) forms a framework for FDA to exempt devices for investigational use. IDE must be submitted and approved by FDA for all clinical evaluations of investigational devices before initiation of the study as described in 21 CFR 812. The clinical and analytical studies for IVDs, which comply with the terms of 812.2(c), are mainly exempted from the IDE requirement. The IDE application requires submission of the following information:

1. An investigational institutional review board (IRB)-approved plan which also needs FDA approval if the study involves a significant risk device
2. Informed consent from all patients
3. Labeling device as investigational use only
4. Monitoring of the study with required documentation for records and reports

Humanitarian Device Exemption (HDE) is another type of FDA submission, which should be made in exceptional cases for humanitarian use device (HUD). The HUD is a device, which is proposed to benefit patients with condition that affects less than 4000 individuals in the United States per year. This can be considered as incentive to the device manufacturer's costs which may exceed its market returns for such small patient populations (US-FDA [online]).

8.2.4 Analytical, Clinical, and Other Regulatory Requirements for Biomarker Evaluation

There is increasing demand from regulatory bodies for validation of any novel biomarker before their translation in clinical practice, to ensure patient's welfare. The research scientists and lab professionals should be acquainted with these regulatory requirements. This will allow the better use of discovered biomarkers as valid medical tests (or device). Several analytical and clinical requirements including cost-effectiveness and larger impact of biomarker assay tests are expected before moving these devices from bench to bedside. The intended use of a device is of vital importance, which directs all clinical, analytical, and other requirements for biomarker performance. Thus, these requirements can vary substantially depending on intended use of device. As described earlier the classification of devices in classes I, II, and III is of primary consideration before deriving framework for evaluation tests (Anderson and Anderson 2002; Ong and Mann 2005; Horvath et al. 2014).

8.2.4.1 Intended Use (Purpose of the Test)

It is of paramount importance to understand the intended use of the device or biomarker assay, as it will determine its overall clinical performance, safety, and other evaluation criteria. This describes the intended clinical application and substantial improvement in medical practice of the test. The intended use may include diagnosis, risk assessment, screening, therapeutic monitoring, or treatment. It is equally important to recognize that the test is a part of medical treatment, and it should be well integrated into clinical management system to improvise the overall diagnostic or treatment with reduced cost. Defining the test purpose can be the most critical part of the application process. A separate guideline issued by FDA on "Determination of Intended Use for 510(k) Devices" can be referred in this context.

8.2.4.2 Device Description

The device for in vitro diagnostic tests consists of all the items of a test system such as reagents, instruments including software, and data analysis tools. The regulatory application should define all the components of assay system with their principle and protocols.

The quality system evaluation as per 21CFR 820.50 requires establishing and controlling procedures to ensure quality of all test items as per set specifications. In case of relatively new mass spectrometry-based proteomic assays, quality should be established against existing methods such as ELISA and immunoassays. For MS-based proteomic assays, test system may include one or more sample preprocessing, liquid chromatography and mass spectrometry components, raw data acquisition and storage, data analysis software, etc. The FDA requires establishing validity and reliability of any technology used for assay development (Lee et al. 2011; Li et al. 2011; Lopez et al. 2012; Haleem and Timothy 2013).

8.2.4.3 Analytical Performance

Discovery of promising biomarker candidate should lead to the development of a diagnostic assay with reliable analytical performance, thereby validating its accuracy, precision, and quantification. The discovery proteomic tools such as MS-based qualitative methods are normally not well validated and therefore are not suitable for clinical applications. Hence, alternative or modified methodologies such as targeted proteomics can be considered for validation as per following parameters (Anderson and Anderson 2002; Ong and Mann 2005; Wood et al. 2013; Horvath et al. 2014).

Precision

It indicates the degree of agreement between a series of measurements for the particular test sample under defined experimental conditions. Precision evaluates standard deviations or errors for intraday, inter-day, and interlaboratory measurements. These are in general considered as repeatability (intra- and inter-day) and reproducibility (interlaboratory) of the method, which are primary requirements of good analytical performance. Precision is quantitatively reported in terms of the standard deviation (SD), relative standard deviation (RSD), or coefficient of variation (CV) for repeated measurements. Precision often varies with analyte concentrations measured, and hence the precision tests should be performed in a series of concentrations from lower to higher levels.

Several guidelines define acceptable limits of precision in terms of maximum value of CV or SD at given analyte concentrations. However, the intended use of method in clinical practice is the major determining factor in defining the precision acceptance criteria.

Accuracy

The closeness of agreement between the measured value and a true or reference value is accuracy. It evaluates systematic error in measurement and is quantitatively expressed in terms of deviation from reference or known value. The degree of error in accuracy is critical and often the foremost criteria for reliability of test procedures and hence such bias should be eliminated during the biomarker discovery and translational stages.

Limit of Detection (LOD) and Limit of Quantitation (LOQ)

It is the lowest concentration of an analyte that can be detected (LOD) or precisely and accurately determined (LOQ) using the defined analytical method or test procedure. This defines the sensitivity of biomarker measurement and is an important factor in deciding suitability of analytical method for intended clinical use.

Specificity

The analytical interference arising from biological constituents of sample other than analyte (matrix) is a common phenomenon for complex biological samples such as blood plasma, tissues, etc. The presence of complex matrix can suppress (matrix effect) or give false-positive measurements.

The analytical method must provide acceptable specificity for targeted biomarker with required sensitivity by eliminating matrix effect. It is advisable to establish the method specification and selectivity at biomarker discovery and translational stage.

8.2.4.4 Clinical Performance

Reliable and robust analytical performance is indispensable for qualification of clinical biomarker assay development. But the biomarker will fail the evaluation if it is not providing intended clinical performance. The intended use, clinical utility, clinical sensitivity, and specificity need to be considered while assessing the clinical performance of the biomarker assay.

8.2.4.5 Other Additional Considerations

In addition to the above aspects, there are several practical issues, which need to be considered as performance of assay method. The speed of analysis and efficiency (high throughput) of analytical methods can be important criteria for modern commercial laboratories. Also high-throughput results are often necessary during medical procedures where turnaround time is critical in decision making for physicians and clinicians. The cost-effectiveness and affordability of assay are vital consideration in terms of taking the healthcare service to larger communities in the developing world (Lee et al. 2011).

Figure 8.1 briefly describes the biomarker development cycle from discovery to regulatory approval and product launch.

8.2.5 Proteomic Technologies for Biomarker Research

Here we briefly describe the mass spectrometry-based methodologies available for proteomic biomarker research. More details on different technologies for biomarkers discovery have been described in Chaps. 3 and 4 of this book.

8.2.5.1 Mass Spectrometry-Based Proteomics

Mass spectrometry has now become a primary tool for profiling and quantification of proteins. Many discovery and targeted quantitative proteomic approaches have proven their potential for the discovery of novel biomarkers for disease diagnosis and therapeutic drug monitoring (Li et al. 2011).

8.2.5.2 Targeted Proteomics

The biomarker discovery process or methodology may not be a concern for regulatory bodies. However, the most critical part is translating such proteomic biomarkers from discovery phase to medical test, for clinical practice.

The discovery-based approach is often not suitable for routine laboratory testings due to lack of a comprehensive validation workflow. The quantitative ability of such mass spectrometry-based methods needs to be validated for accuracy, precision, linearity, limit of detection (LOD), Limit of Quantitation (LOQ), matrix effect, stability, and robustness. A targeted mass spectrometry method utilizes liquid chromatography and triple quadruple mass spectrometry (LC-MS/MS) settings in multiple reaction monitoring (MRM) mode as effective alternative to immunoassays for quantitative measurements.

Targeted LC-MS/MS approach relies on measuring ion intensities of specific fragment (MS/MS, product) ions for selected precursor (parent) ion of targeted analyte (here peptide). The targeted peptides can be quantified by comparing relative peak areas with other sets of experiment conditions or appropriate labeled internal standard (IS). For absolute quantification calibration curves were plotted using standards of known concentration across linear range of target concentration while considering dynamic range of mass spectrometer. Due to their ability to select and fragment specific target ions, the triple quadruple mass spectrometers gain advantage in terms of sensitivity, specificity, and dynamic range of detection. As mentioned earlier, the analytical performance criteria are of great importance in translating targeted proteomic methods in clinical practice.

8.2.5.3 Discovery Proteomics

Unlike targeted proteomics, discovery proteomics deals with profiling of mass spectrometry-based signatures of two different experimental sets, with or without knowing the identity of proteins or peptides.

This analysis utilizes high-resolution mass spectrometry-based techniques such as matrix-assisted laser desorption/ionization (MALDI) or electrospray ionization (ESI) with time-of-flight (TOF) or orbitrap mass analyzers. This type of analysis is generally a research based and aims at identifying differential expressed set of proteins using mass spectrometry signals. The early stage of analysis is much exploratory in nature to discover marker proteins and not necessarily well validated. Such methods may not be suitable for translating into clinical practice unless validated using targeted proteomic approach and usually are not of concern to regulatory bodies. However, it is important to establish the accuracy and specificity of biomarkers at this stage to facilitate its smooth translation into diagnostic settings.

8.2.6 Overview of Patents for Biomarkers

The journey of a biomarker from the discovery phase is bifurcated during its path to translation where one path ensures that a biomarker meets the global regulatory guidelines, while the other focuses on protecting the intellectual property rights of the researcher who has invested his resources into developing a product which could be potentially used for public welfare. It is therefore of paramount importance to successfully commercialize biomarkers in order to ensure sufficient financial support for biomarker research. Protection of the commercial biomarkers can be achieved through patents; but most often leads to complex intellectual property (IP) issues owing to the fact that ownership of such biomarkers usually involves multiple collaborators. Nevertheless, over the last few years, several patented biomarkers are being developed and validated in large-scale clinical trials.

A patent is an exclusive right granted to an inventor for a defined period of time for an invention, which is novel and nonobvious. In return, the patent holder is expected to disclose to the world and complete details of his invention. The patent thus granted provides the inventor the right to make and sell his invention, simultaneously forbidding the use of the work by others, unless permitted through licensing. The US patent law states that "any new and useful process, machine, manufacture, composition of matter or improvement thereof" is patent eligible, provided it is novel (i.e., not previously known or used in the United States or described anywhere) (Offit et al. 2013, U.S Patent law [Online]). Inventions that preempt the laws of nature or hypothetical ideas and physical phenomena, termed as judicial exceptions, are considered patent ineligible since they do not meet these patentability requirements (Andrews et al. 2006). In such situations, protection can be obtained as a trade secret that does not require any information to be disclosed and is valid for an indefinite period of time. However, it should be noted that it does not give the inventor an exclusive right over his invention and cannot prevent third parties from commercializing the work upon discovering the confidential information. The situation becomes slightly more complicated and challenging while issuing patents for biomarker-related inventions.

8.2.7 Biomarker Patent Claims

The power of patent protection depends on the type of biomarker discovery. Biomarkers that provide information on a patient's response to a particular therapy are more likely to involve a known marker or a group of known markers. In such cases the novelty of the claim is the mere association of these biomarkers with a particular drug. Such correlation claims involve patent eligibility issues as the association itself is likely to be considered a "natural principle."

Biomarker patent claims can be subsumed into two broad categories:

Composition of matter: Patent claims of this nature provide the strongest type of protection for novel biomarkers such as metabolites, proteins, genes, and gene expression products which are significantly different from naturally occurring substances. Protection of such biomarkers that, in most cases, serve as key players in disease diagnosis prohibits their use by other parties for commercial purposes. What is of paramount importance here is the level of uncertainty these claims carry with them with respect to "patentability." This is due to the extremely thin line that distinguishes these biomarkers from products of nature. Most of them often qualify as "unpatentable subject matter" owing to the fact that the inventors fail to provide sufficient evidence to prove the novelty of the nonnatural composition of the "inventive step" that the claim recites.

A few examples of "composition of matter" claims include patents issued to natural metabolites isolated from *E. turbinate* such as ecteinascidins, exhibiting antibacterial activities as well as antitumor activities (Rinehart and Holt 1992). These metabolites along with the products of their metabolism were patent protected with the intention of utilizing them as medicinal agents to inhibit the growth of cancer cell (Rinehart et al. 2001). In the year 1995, a patent was granted for the production of recombinant erythropoietin (Lin 1995). Erythropoietin, a hormone that controls the production of red blood cells, finds tremendous application in diagnosis and treatment of blood diseases. Several methods for purifying and stabilizing human erythropoietin have also generated enormous number of patents (Schlueter and White 1962; Chiba et al. 1975; Egrie 1985). Therapeutic powders containing human proteins such as insulin along with enhancers for increased absorption into the system when administered to patients with diabetes (Jensen and Hansen 2000) have also been patented. Recently in 2010, Intrinsic Bioprobes Inc. (IBI) was granted a patent by USPTO on novel blood biomarkers and novel assays used for detection of a panel of these specific human protein variants with clinical applications in the diagnosis and monitoring of cardiovascular diseases (Caulfield et al. 2000; Merz and Cho 2005; Kiernan et al. 2010). While on one hand, these patents offer protection of the highest order, an overriding concern accompanying them is the level of hindrance they cause to basic scientific research. Let us understand this through a case study on the patenting of human genes.

The *Myriad Genetics* case: The patent eligibility of human genes has raised enormous concern across the globe for several decades, especially among the molecular diagnostic companies, laboratories, patients, and pharmaceutical industries (Caulfield et al. 2000; Merz and Cho 2005). For instance, let us consider that Company X files a patent for a particular gene. The fundamental discovery in this case is the association of certain mutations in this gene with a disease Y. The patent thus granted provides the patent holder with an exclusive right to use this gene for diagnostic purposes and includes any assay developed to test for this genetic difference. This creates a monopoly since the patent precludes the use of the gene by other companies, permitting Company X to completely capture the diagnostic testing market. This type of patent implies a single testing method even if it is not the most suitable one for diagnosis of disease Y.

The debate on the patentability of genes made headlines when the US court on June 13, 2013, invalidated gene patents that Myriad Genetics Inc. held for more than a decade on the basis that the patent encompassed products of nature. This unanimous decision upended the long-standing precedent that isolated genes were indeed patentable subject matter. In 1994, the molecular diagnostic company cloned the gene BRCA1 followed by BRCA2 associated with breast cancer (Nat Med Editorial, 2013 [Online]). The patent granted by the US court in 1999 gave the American company an exclusive right over the sequence of the BRCA1 gene and the right to test for polymorphisms in the genes responsible for most of the breast cancers. In 2009, a lawsuit was filed by the American Civil Liberties Union and Public Patent Foundation after the EPO issued patents to Myriad Genetics for the genes in Europe (Benowitz 2003). This patent was followed by a third one for the exclusive use of the gene and the protein for therapeutic purposes. The situation demanded that all samples across the world be sent to the company for testing and forbade other laboratories from performing the test or developing alternative diagnostic tests. The exorbitant costs involved as well as the lack of options available to the patients for accurate diagnosis of breast cancer encouraged plaintiffs to challenge the validity of the Myriad's patent claims. The court's decision to reject the patent claims was based on the fact that that neither the genetic information nor gene structure had been altered by Myriad and hence was not any different from the naturally existing DNA molecule. However, the court also declared that cDNA is patent eligible because it differs from DNA in that it does not contain introns, widening the huge gap of uncertainty between what can and cannot be patented. In this light, 8,000 patents involving genes and of these around 3,500 involving human genes may be invalidated (Graff et al. 2013). Then the number of US patent applications involving human genes has also declined ever since.

In spite of these hurdles, the fact remains that a DNA molecule can still be easily engineered in a manner that makes it markedly different from an already existing one in order to be a patentable subject matter. However what is worrisome is a situation in the post-genomic era where genetic material may not be the key biomarkers for diagnosis and therapeutics. For instance, can an inventor claim a patent for the discovery of certain antibiotics and naturally occurring drugs for their use in treating a particular disease or will it also be considered as a prior art by the court? (Harrison 2013). Another concern is whether such laws would indelibly harm the biotechnology sector by lowering the impetus for development of new diagnostic tests.

Method claims: Patent claims of this nature include requests to patent novel methods for assaying diagnostic, prognostic, and pharmacogenomic biomarkers. Genome-wide markers that yield information regarding the response of a patient to therapy are called pharmacogenomic markers. These are extremely important markers to determine whether a particular individual would benefit upon administering a therapeutic agent or not, since the effects of drugs vary greatly among patients (Luo 2012). Based on their genome profiles, personalized therapies most beneficial to the individuals can be designed (Chen et al. 2013). The jurisprudence of countries across the world differs significantly with respect to issuing patents in such cases. For instance, the United States Patent and Trade Office (USPTO) may consider a biomarker assay patent eligible, as against the European Patent Office (EPO), granting a patent solely to the novel device used for the application and not the method itself (Hu and Wong 2014; Andrews 2014 [Online]). Some examples include patents that have been previously issued for the protection of the methods/assays developed for the detection of salivary proteins that serve as biomarkers for distinguishing oral cancer from periodontal disease and for the early diagnosis and prognosis of the former (Hu and Wong 2014). Similarly methods for the detection of single nucleotide polymorphisms associated with the progression of glaucoma and for estimating a high progressive risk of glaucoma (Kinoshita et al. 2013) have also been patented. In some cases biomarker patents are also granted to compounds such as antibodies used for detection of biomarkers (Fisher 2012). In 2013, a patent was granted to methods developed for the measurement of the expression levels of TGFBR1 patients in order to assess their genetic susceptibility of patients to colorectal cancer. The patent claim also included appropriate treatment to the patients upon detection of cancer through colonoscopy (Pasche 2013). Figure 8.2 depicts the significance of patents in biomarker research.

8.2.8 Types of Biomarker Patents

The completion of the human genome project and the rapid advancements in biomarker research have generated the prospect of personalized medicine. The adaptation of these advancements has led several researchers and companies to claim patents on their discovery of biomarkers for both infectious and noninfectious diseases, which can be broadly categorized, based on their clinical applications. Tables 8.3 and 8.4 summarize a few patents granted to biomarkers associated with infectious and noninfectious diseases.

8.2.8.1 Diagnostic Biomarker Patents

The ST2-based diagnosis for dengue fever virus is one such example for application of a biomarker for rapid detection of an infectious disease. Infections caused by the dengue fever

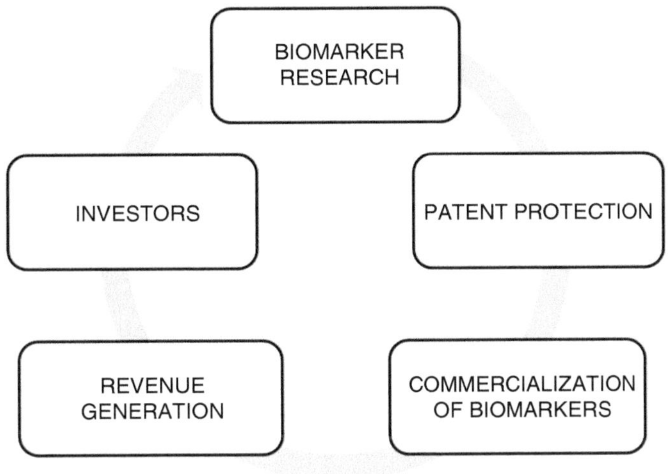

Fig. 8.2 Significance of patent protection in biomarker research

virus have been linked to illnesses stretching from dengue fever, to dengue hemorrhagic fever, to dengue septic shock. Herein, the soluble ST2 protein is understood to act as a biomarker for infections triggered by the dengue fever virus and may be applicable as a diagnostic kit (Bosch and Becerra 2009). In the year 2011, a patent was filed proposing the application of the chemokine receptor 6 (CCR6) as a potential biomarker for Alzheimer's disease. The increment in the expression of CCR6 beyond the threshold in a sample signifies the patient has or is highly prone to the development of Alzheimer's disease (Offner-Vandenbark et al. 2013). In 2011, CIP2A was suggested as a biomarker detected in the cervix of a patient burdened by cervical cancer. Given the strong correlation between aberrant CIP2A expression and cervical cancer, it was proposed as a marker for early detection of cervical cancer in an individual with high sensitivity as well as specificity (Huang and Trama 2012).

8.2.8.2 Prognostic Biomarker Patents

Prognostic biomarkers are the indicative means of measurement of the long-term outcome of patients. They also provide an estimate of the severity and the likely outcome of the disease. Upon diagnosis of a disease, determination of the correct prognosis is essential in order to provide effective disease management insights for a patient. The application of prognostic biomarkers for predicting the severity of an infectious disease like malaria is of grave importance. A patent filed in 2012 highlights the application of CXCL4, CXCL10, VEGF, PGDF, IL-1Ra, IL-8, MIP-1β, sFas, Fas-L, sTNF-R2, and sTNF-R1 as outlining the severity of malaria and predicting mortality due to cerebral malaria (Stiles et al. 2013).

In 2013, a patent was filed to provide gene expression profiles indicative of whether a patient plagued with progesterone receptor compelled malignancies is expected to respond to treatment with an antiprogestin therapeutic compound. Identification of such responsiveness makes it easier for a treatment provider to determine in advance the patients who may benefit from such treatment as well as identify an alternative therapeutic route for the nonresponders (Lange et al. 2013).

8.3 Conclusion

This chapter summarizes various regulatory guidelines obligatory for biomarker research and its translation to commercialize in diagnosis and prognosis of various diseases. Also a brief compilation of the most important regulatory bodies worldwide has been included to provide the readers an insight to industrial regulatory affairs involved in drug development. This chapter gives overview of regulatory perspectives particularly

Table 8.3 List of diagnostic and prognostic biomarkers for infectious diseases

S. No	Infectious disease	International publication number	Biomarker	Nature of the biomarker	Effect
1.	Anthrax	US 8,591,899 B2(Shafferman et al. 2013)	BA 3660	Serine protease	Diagnosis of anthrax infection
			BA1952	Endopeptidase	
			BA 0796	Hypothetical protein	
2.	Dengue	WO 2009/145810 A2 (Bosch and Becerra 2009)	ST2	Cytokine receptor family member	Diagnosis of dengue fever virus infection
		WO 2013/130029 A1 (Hunsperger et al. 2013)	Vitronectin	Glycoprotein	Diagnosis, methods for determining severity of infection and differentiation of dengue fever and dengue hemorrhagic fever/dengue shock syndrome
3.	HIV	US 8,273,538 B2 (Rappaport and Fischer-Smith 2012)	CD163	Protein that marks cells of monocyte/macrophage lineage that is a scavenger receptor for the hemoglobin-haptoglobin complex	Monitoring the course of HIV-1 infection and/or HIV encephalopathy and predicting viral load and changes in viral load. May also play a role in diagnosis or risk assessment relative to CNS and other neurological manifestations of HIV infection
4.	Lyme disease	WO 2012/040264 A2 (Costello et al. 2012)	Autoantigen ECGF	Platelet-derived endothelial cell growth factor	Diagnosis of Lyme disease-mediated arthritis
5.	Malaria	US 2013/0108647 A1 (Stiles et al. 2013).	CXCL4	Also known as platelet factor 4, is a small cytokine that belongs to the chemokine family	Diagnosis, prognosis, and detecting severity of malaria
			CXCL10	Also known as IP-10, is a small cytokine that belongs to the chemokine family	
			VEGF	Signal proteins produced by the cells that stimulate vasculogenesis and angiogenesis	
			PGDF		
			IL-1Ra	Cytokine receptor that binds to interleukin-1	
			IL-8	Chemokine of the immune system	
			MIP-1β	Also known as CCL4, is a chemokine with specificity for CCR5 receptor	
			sFas	Receptor for Fas-L, involved in apoptosis	
			Fas-L	Transmembrane protein that belongs to TNF family involved in apoptosis	
			sTNF-R2	A member of the TNF receptor-associated factor protein family that mediates signal transduction	
			sTNF-R1	Same as sTNF-R2	

(continued)

Table 8.3 (continued)

S. No	Infectious disease	International publication number	Biomarker	Nature of the biomarker	Effect
7.	Pneumonia	US 2013/0296240 A1 (Pemberton et al. 2013)	GHRsp fragment biomarkers	Ghrelin signal peptide	Diagnosis, prognosis, risk stratification, assessing, staging, monitoring, categorizing, and determination of further diagnoses and treatment regimens in subjects with pneumonia
8.	Urinary tract infection	WO 2012/107450 A1 (Åkesson et al. 2012)	HBP (heparin-binding protein)	Serine protease that is present within the secretory vesicles of human neutrophils	Diagnosis and treatment of urinary tract infection

Table 8.4 List of diagnostic and prognostic biomarkers for other diseases

S. No	Diseases	International publication number (Ref)	Biomarker	Nature of the biomarker	Effect
1	Alzheimer's disease (AD)	WO/2013/176885 A1 (Todd and Thuc 2013)	sTNF-R2	Tumor necrosis factor receptor 2	Early diagnosis, determination of severity and treatment of AD
			Abeta-42	Amyloid beta protein	
			PTau-181	Phosphorylated forms of Tau	
			IL-2	Interleukins	
			Abeta oligomers	Oligomers	
		US 2013/0316338 A1 (Offner-Vandenbark et al. 2013)	CCR6	Chemokine receptor	Diagnosis
2	Breast cancer	US 2013/0316992 A1 (Lange et al. 2013)	Antiprogestin responder gene signatures	Genes	Prognosis based on gene expression profiles of a patient
3	Cervical cancer	US 2012/0070837 A1 (Huang and Trama 2012)	CIP2A	mRNA	Detection of cervical cancer
			Ki67	Nuclear protein associated with cell proliferation	
			TOP2A	Cell cycle regulation genes	
			MCM2	Mini chromosome maintenance protein	
			MCM5	Mini chromosome maintenance protein	
			p14, p16	Cell cycle regulation genes	
4	Colorectal cancer	WO/2013/066641 A1(Schutz et al. 2013)	CNA	Circulating nucleic acids	Diagnosis
5	Cardiovascular disorders	US 2013/0316378 A1 (Pemberton et al. 2013)	BNP-SP	Signal peptide	Prediction, diagnosis, and monitoring of acute cardiac disorders

for scientist in biomarker discovery research community. Patents of many infectious as well as noninfectious diseases have been discussed in the chapter to provide extended information on the intellectual property rights of the existing biomarkers of those that have been successfully discovered and used.

Competing Interests The authors declare no competing interests.

References

A victory for genes (2013) Nat Med 19:792–792. doi:10.1038/nm.3279

Åkesson P, Björck L, Christensson B et al (2012) Diagnostic method for urinary tract infection. WO2012107450, 16 Aug 2012

Anderson NL, Anderson NG (2002) The human plasma proteome: history, character, and diagnostic prospects. Mol Cell Proteomics MCP 1:845–867

Andrews HD (2014) Strategies for protecting biomarker innovation in the US and abroad, 19 Feb 2014. http://baybio.org/lunch-learn-strategies-protecting-biomarker-innovation-us-abroad/

Andrews L, Paradise J, Holbrook T, Bochneak D (2006) When patents threaten. Science 314:1395–1396. doi:10.1126/science.1135872

Benowitz S (2003) European groups oppose Myriad's latest patent on BRCA1. J Natl Cancer Inst 95:8–9

Biologicals, Central drugs standard control organization, New Delhi (July 2015 Update). http://www.cdsco.nic.in/forms/list.aspx?lid=1812&Id=1

Bosch I, Becerra A (2009) St2-based dengue fever diagnostic. WO2009145810, 3 Dec 2009

Caulfield T, Gold ER, Cho MK (2000) Patenting human genetic material: refocusing the debate. Nat Rev Genet 1:227–231. doi:10.1038/35042087

Chen JJ, Lin W-J, Chen H-C (2013) Pharmacogenomic biomarkers for personalized medicine. Pharmacogenomics 14:969–980. doi:10.2217/pgs.13.75

Chiba S, Kung CKH, Goldwasser E (1975) Stabilization of urinary erythropoietin using sodium p-aminosalicylate and extracting into phenol. US Patent 3,865,801, 11 Feb, 1975

Consolidated Patent Laws (May 2015 Update) United States Code Title 35 – Patents (2015) http://www.uspto.gov/web/offices/pac/mpep/consolidated_laws.pdf

Costello C, Drouin E, Seward R, Steere AC (2012) Diagnostic tests for immune reactivity with human endothelial cell growth factor. WO2012040264, 29 Mar, 2012

Draft Guidance for Industry and FDA Staff Medical Devices (Jan 2015 Update). http://www.fda.gov/MedicalDevices/DeviceRegulationandGuidance/GuidanceDocuments/ucm418408.htm

E15 (Definition for genomic biomarkers, pharmacogenomics, pharmacogenetics, genomic data and sample coding categories) (2007) ICH guidelines. http://www.ich.org/fileadmin/Public_Web_Site/ICH_Products/Guidelines/Efficacy/E15/Step4/E15_Guideline.pdf

E16 (Biomarkers related to drug or Biotechnology product development: context, structure, and format of qualification submissions) (2010) ICH guidelines. http://www.ich.org/fileadmin/Public_Web_Site/ICH_Products/Guidelines/Efficacy/E16/Step4/E16_Step_4.pdf

Egrie JC (1985) Immunoreactive with polypeptide sequence extant of erythropoietin; detecting red blood cell growth stimulating hormone. US Patent 4,558,006, 10 Dec 1985

Fisher D (2012) Biomarker patents: where do we stand now? Biotechnol. Focus

Füzéry AK, Levin J, Chan MM, Chan DW (2013) Translation of proteomic biomarkers into FDA approved cancer diagnostics: issues and challenges. Clin Proteomics 10:13. doi:10.1186/1559-0275-10-13

Government and regulatory bodies, Pharmaweb. http://www.pharmweb.net/regulatory.html

Graff GD, Phillips D, Lei Z et al (2013) Not quite a myriad of gene patents. Nat Biotechnol 31:404–410. doi:10.1038/nbt.2568

Haleem, Timothy (2013) Proteomic and metabolomic approaches to biomarker discovery. Elsevier Inc. ISBN: 978-0-12-394446-7

Harrison C (2013) Isolated DNA patent ban creates muddy waters for biomarkers and natural products. Nat Rev Drug Discov 12:570–571. doi:10.1038/nrd4084

Horvath AR, Lord SJ, StJohn A et al (2014) From biomarkers to medical tests: the changing landscape of test evaluation. Clin Chim Acta Int J Clin Chem 427:49–57. doi:10.1016/j.cca.2013.09.018

Hu S, Wong DTW (2014) Salivary protein biomarkers for human oral cancer. US Patent 8,728,744, 20 May 2014

Huang LP, Trama J (2012) CIP2A as a biomarker in detection of cervical cancer. US Patent 20120070837, 22 Mar 2012

Hunsperger E, Ndao M, Tomashek K, Poole-Smith BK (2013) Detection of subject biomarker diagnostic assay for dengue fever and the differentiation of dengue hemorrhagic fever.WO2013130029, 6 Sept 2013

ICH official website, work products, Geneva. http://www.ich.org/products/guidelines.html

Jensen S, Hansen P (2000) Method for producing powder formulation comprising an insulin. US Patent 6,043,214, 28 Mar 2000

Kiernan UA, Nedelkov D, Nelson RW (2010) Biomarkers and assays for myocardial infarction.US Patent 7,816,095, 19 Oct 2010

Kinoshita S, Tashiro K, Nakano M et al (2013) Method for determination of progression risk of glaucoma

Lange C, Knutson T, Nikas JB (2013) Breast cancer prognosis.US Patent 20130316992, 28 Nov 2013

Lee J, Han JJ, Altwerger G, Kohn EC (2011) Proteomics and biomarkers in clinical trials for drug development. J Proteomics 74:2632–2641. doi:10.1016/j.jprot.2011.04.023

Li J, Kelm KB, Tezak Z (2011) Regulatory perspective on translating proteomic biomarkers to clinical diagnostics. J Proteomics 74:2682–2690. doi:10.1016/j.jprot.2011.07.028

Lin F-K (1995) Production of recombinant erythropoietin. US Patent 5,441,868, 15 Aug 1995

López E, Madero L, López-Pascual J et al (2012) Clinical proteomics and OMICS clues useful in translational medicine research. Proteome Sci 10:35

Luo W (2012) Method for discovering pharmacogenomic biomarkers. WO 2012106267, 9 Aug, 2012

Merz JF, Cho MK (2005) What are gene patents and why are people worried about them? Community Genet 8:203–208. doi:10.1159/000087956

Nagasaki S, Lang B, Shintani M et al (2008) An overview of pharmaceutical and medical device regulation in Japan. Special report life sciences. Asian counsel

Offit K, Bradbury A, Storm C et al (2013) Gene patents and personalized cancer care: impact of the Myriad case on clinical oncology. J Clin Oncol Off J Am Soc Clin Oncol 31:2743–2748. doi:10.1200/JCO.2013.49.7388

Offner-Vandenbark H, Quinn JF, Subramanian S (2013) CCR6 As a biomarker of Alzheimer's disease. US Patent 20130316338, 28 Nov 2013

Ong S-E, Mann M (2005) Mass spectrometry-based proteomics turns quantitative. Nat Chem Biol 1:252–262. doi:10.1038/nchembio736

Outline of PMDA, pharmaceuticals and medical devices agency, Tokyo, Japan. http://www.pmda.go.jp/english/about/

Pasche B (2013) TGFBR1 expression modifies risk for colorectal cancer. US Patent 8,492,096, 23 Jul, 2013

Pemberton CJ, Richards AM, Byers MS (2013) Pneumonia biomarkers. US Patent 20130296240, 7 Nov 2013

Pemberton CJ, Richards AM, Nicholls MG, Yandle TG (2013) Methods of diagnosing acute cardiac disorders using bnp-sp. US Patent 20130316378, 28 Nov 2013

Pharmaceutical development, European Medicines Agency, London. http://www.ema.europa.eu/ema/index.jsp?curl=pages/regulation/general/general_content_000362.jsp&

Rappaport J, Fischer-Smith T (2012) Blood monocyte CD163 expression as a biomarker in HIV-1 infection and neuroaids. US Patent 8,273,538, 25 Sep 2012

Rinehart KL, Holt TG (1992) Ecteinascidins 729, 743, 745, 759A, 759B and 770. US Patent 5,089,273, 18 Feb 1992

Rinehart KL, Morales JJ, Reid J, Reymundo I, Floriano P, Gravalos LG (2001) ETM-775 metabolite of ecteinascidin 743. US Patent 6,316,214, 13 Nov 2001

Schutz E, Beck J, Urnovitz H (2013) Colorectal cancer associated circulating nucleic acid biomarkers. WO2013066641, 10 May 2013

Shafferman A, Cohen O, Chitlaru T et al (2013) Diagnosis of Bacillus anthracis infection based on detection of bacterial secreted biomarkers. US Patent 8,591,899, 26 Nov 2013

Shreeve RL, Banks RE, Selby PJ, Vasudev NS (2012) Proteomic study of conditioned media: cancer biomarker discovery. Int J Genomics Proteomics 3(1):50–56

Schlueter RJ, White WF (1962) Erythropoietic factor purification. US Patent 3,033,753, 8 May 1962

Stiles JK, Lillard JW, Armah HBA et al (2013) Compositions and methods for diagnosis, prognosis and management of malaria. US 2013/0108647 A1, 2 May 2013

Thalidomide crisis and drug regulation, Emory Library Information Technology, Atlanta (July 2015 Update). http://guides.main.library.emory.edu/disasterdisplay?p=325039

Todd JA, Thuc, LQA (2013) Biomarkers for Alzheimer's disease. WO2013176885, 28 Nov 2013

U.S. Food and Drug Administration, official website, New Hampshire (July 2015 Update). http://www.fda.gov/ForIndustry/default.htm

Vidal M, Chan DW, Gerstein M et al (2012) The human proteome – a scientific opportunity for transforming diagnostics, therapeutics, and healthcare. Clin Proteomics 9:6. doi:10.1186/1559-0275-9-6

Witcher MF, Odum J (2012) Biopharmaceutical manufacturing in the twenty-first century- the next generation manufacturing facility. Pharm Eng 32(2)

Wood SL, Knowles MA, Thompson D et al (2013) Proteomic studies of urinary biomarkers for prostate, bladder and kidney cancers. Nat Rev Urol 10:206–218. doi:10.1038/nrurol.2013.24

Index

S. Srivastava (ed.), *Biomarker Discovery in the Developing World: Dissecting the Pipeline for Meeting the Challenges*, DOI 10.1007/978-81-322-2837-0